Disclaimer

The publisher of this book is by no way associated with the National Institute of Standards and Technology (NIST). The NIST did not publish this book. It was published by 50 page publications under the public domain license.

50 Page Publications.

Book Title: An Ontology for the e-Kanban Business Process

Book Author: Edward J. Barkmeyer; Boonserm Kulvatunyou

Book Abstract: Large automotive manufacturers, including automakers and the manufacturers of principal automotive subsystems, make their products in large volumes. This means that the demands on their suppliers are fairly predictable over a long term, but subject to local peaks and valleys that result from variations in short-term demand and production schedules. As a consequence, the industry has found it expedient to develop vendor-managed inventory arrangements with many of the suppliers of commonly used parts and materials. In such a scheme, the principal manufacturer maintains only a few days or weeks inventory of the parts and keeps the supplier informed of the actual rate of consumption, and the supplier arranges to deliver parts and materials just in time to maintain the inventory level needed for immediate manufacture. This reduces the cost of space, time, personnel and equipment for maintaining the parts and materials inventory at the manufacturing facility. At the same time, it gives the supplier better information and better control over the replenishment process, which allows him to plan his production to a known customer demand.

Citation: NIST Interagency/Internal Report (NISTIR) - 7404

Keyword: business process;inventory;manufacturing;ontology;production process

NISTIR 7404

An Ontology for the e-Kanban Business Process

Edward J. Barkmeyer
Boonserm Kulvatunyou

Manufacturing Engineering Laboratory,
National Institute of Standards and Technology

Table of Contents

1 Introduction 1
2 Basic Model Elements 5
    2.1 Datatypes 5
        2.1.1 Datatype: Boolean 5
        2.1.2 Datatype: CodeType 5
        2.1.3 Datatype: DateTime 6
        2.1.4 Datatype: DecimalNumber 6
        2.1.5 Datatype: Integer 7
        2.1.6 Datatype: Name 7
        2.1.7 Datatype: Quantity 7
        2.1.8 Datatype: Text 8
        2.1.9 Datatype: TimePeriod 8
        2.1.10 Datatype: URI 9
        2.1.11 Datatype: UnitOfMeasure 9
    2.2 Classes 9
    2.3 Instances 9
3 Processes and Roles 10
    3.1 Datatypes 11
        3.1.1 Datatype: RoleName 11
    3.2 Classes 11
        3.2.1 Class: BusinessProcess 11
        3.2.2 Class: CustomerParty 12
        3.2.3 Class: E-Kanban 13
        3.2.4 Class: E-KanbanPartyRole 13
        3.2.5 Class: PartyRole 14
        3.2.6 Class: ProcessSpecification 15
        3.2.7 Class: RoleSpecification 16
        3.2.8 Class: SupplierParty 16
    3.3 Instances 17
4 Identifiers 18
    4.1 Datatypes 18
        4.1.1 Datatype: AgencyId 18
    4.2 Classes 19
        4.2.1 Class: CustomerId 19
        4.2.2 Class: ID 19
        4.2.3 Class: PartyOwnedRegistry 20
        4.2.4 Class: Registry 20
        4.2.5 Class: SupplierId 21
    4.3 Instances 21
        4.3.1 DUNS 21
5 Parties 22
    5.1 Datatypes 22
        5.1.1 Datatype: Communication 22
        5.1.2 Datatype: Email 23
        5.1.3 Datatype: TelephoneNumber 23
    5.2 Classes 24
        5.2.1 Class: Contact 24
        5.2.2 Class: DUNSID 24
        5.2.3 Class: Facility 25
        5.2.4 Class: Party 25
        5.2.5 Class: PartyId 26

5.3	Instances		27
6	Documents		28
6.1	Datatypes		28
	6.1.1	Datatype: DocumentTypeCode	28
	6.1.2	Datatype: LanguageCode	29
6.2	Classes		30
	6.2.1	Class: Description	30
	6.2.2	Class: Document	30
	6.2.3	Class: DocumentId	31
	6.2.4	Class: DocumentReference	32
	6.2.5	Class: Note	33
6.3	Instances		34
7	Locations		35
7.1	Datatypes		35
	7.1.1	Datatype: LocationId	35
	7.1.2	Datatype: LocationTypeCode	36
7.2	Classes		36
	7.2.1	Class: Location	36
	7.2.2	Class: LocationReference	37
7.3	Instances		38
8	Items		39
8.1	Datatypes		39
8.2	Classes		39
	8.2.1	Class: CustomerItemId	39
	8.2.2	Class: Item	40
	8.2.3	Class: ItemId	40
	8.2.4	Class: StandardPack	41
8.3	Instances		42
9	Shipments and Shipment Schedules		43
9.1	Datatypes		43
	9.1.1	Datatype: ContainerTypeCode	43
	9.1.2	Datatype: DataIdentifierCode	44
	9.1.3	Datatype: LabelPositionCode	44
	9.1.4	Datatype: MarksAndNumbersTypeCode	45
	9.1.5	Datatype: MarksAndNumbers	45
	9.1.6	Datatype: ScheduleType	46
9.2	Classes		47
	9.2.1	Class: Container	47
	9.2.2	Class: Package	48
	9.2.3	Class: PackagingLabel	48
	9.2.4	Class: ReceiptDiscrepancy	49
	9.2.5	Class: ScheduleLine	50
	9.2.6	Class: ShipFromParty	53
	9.2.7	Class: Shipment	54
	9.2.8	Class: ShipmentSchedule	57
	9.2.9	Class: ShipmentUnit	59
	9.2.10	Class: ShipToParty	61
9.3	Instances		61
10	Carriers and Equipment		62
10.1	Datatypes		62
	10.1.1	Datatype: EquipmentTypeCode	62
10.2	Classes		63
	10.2.1	Class: CarrierParty	63
	10.2.2	Class: Equipment	64

	10.2.3	Class: EquipmentOwnerParty	64
10.3	Instances		65

11 Kanbans 66
- 11.1 Datatypes 66
  - 11.1.1 Datatype: KanbanStatusCode 66
- 11.2 Classes 67
  - 11.2.1 Class: Kanban 67
  - 11.2.2 Class: KanbanLoop 68
  - 11.2.3 Class: KanbanStatus 69
- 11.3 Instances 69

12 Messages 70
- 12.1 Datatypes 70
  - 12.1.1 Datatype: ActionCode 70
  - 12.1.2 Datatype: ConfirmationCode 71
  - 12.1.3 Datatype: OAGISNoun 71
- 12.2 Classes 72
  - 12.2.1 Class: BusinessObjectDocument 72
  - 12.2.2 Class: Message 72
  - 12.2.3 Class: System 73
  - 12.2.4 Class: SyncKanbanConsumption 75
  - 12.2.5 Class: KanbanConsumption 75
  - 12.2.6 Class: SyncReceiveDelivery 76
  - 12.2.7 Class: ReceiveDeliveryNotification 77
  - 12.2.8 Class: SyncShipment 78
  - 12.2.9 Class: ShipmentNotification 79
  - 12.2.10 Class: SyncShipmentSchedule 80
- 12.3 Instances 81

References 82

Table of Figures

Figure 1	Basic Datatypes	5
Figure 2	Business Processes	10
Figure 3	E-Kanban Process and Roles	10
Figure 4	Registered Identifiers	18
Figure 5	Parties	22
Figure 6	Documents	28
Figure 7	Locations	35
Figure 8	Items	39
Figure 9	Overview of Shipments and Shipment Schedules	43
Figure 10	Schedule Lines	50
Figure 11	Shipments	54
Figure 12	ShipmentSchedules	57
Figure 13	Shipment Units	59
Figure 14	Carriers and Equipment	62
Figure 15	Kanban Concepts	66
Figure 16	Messages	70
Figure 17	SyncKanbanConsumption Message	75
Figure 18	SyncReceiveDelivery Message	77
Figure 19	SyncShipment Message	79
Figure 20	SyncShipmentSchedule Message	80

An Ontology for
the e-Kanban Business Process

1 Introduction

Inventory Visibility

Large automotive manufacturers, including automakers and the manufacturers of principal automotive subsystems, make their products in large volumes. This means that the demands on their suppliers are fairly predictable over a long term, but subject to local "peaks and valleys" that result from variations in short-term demand and production schedules. As a consequence, the industry has found it expedient to develop "vendor-managed inventory" arrangements with many of the suppliers of commonly used parts and materials. In such a scheme, the principal manufacturer maintains only a few days or weeks inventory of the parts and keeps the supplier informed of the actual rate of consumption, and the supplier arranges to deliver parts and materials "just in time" to maintain the inventory level needed for immediate manufacture. This reduces the cost of space, time, personnel and equipment for maintaining the parts and materials inventory at the manufacturing facility. At the same time, it gives the supplier better information and better control over the replenishment process, which allows him to plan his production to a known customer demand.

Unsurprisingly, software systems have been developed to support the information flows required by this process. Importantly, some of these systems, commonly called *Inventory Visibility* systems, are accessed via Web browsers, which makes them available to smaller supplier enterprises that do not themselves have sophisticated software. As a consequence, most vendor-managed inventory information now flows electronically through the Internet and other networks.

On the other hand, this market for Inventory Visibility (IV) systems has led to the development of many competing ones. While the manufacturer usually chooses to interact with a single third-party inventory visibility system, a supplier who has more than one such customer may be required to interact with a different IV system for each customer. This requires the supplier's personnel to use, and be trained to use, multiple IV systems, and to be aware of the differences in the system interfaces and behaviors, as well as the differences in customer requirements. It also requires the small suppliers to take data out of these systems in different ways, so that the data can be entered into their own planning and accounting spreadsheets. For a small supplier, interaction with multiple IV systems can be a significant cost and a source of many errors.

In order to reduce this burden on automotive suppliers, and to facilitate reconfiguration of supplier networks, the Automotive Industry Action Group (AIAG) has begun development of a set of voluntary standards for the Inventory Visibility software products that will allow the exchange of the related information in a standard form among these third-party systems. The idea is to allow a small supplier to use a single IV system for exchange of information with all of his customers, thus minimizing his training and usage costs. As now, the supplier's IV system would communicate directly with those of his customers who use the same system, but the standard would allow it to communicate the information with other IV systems used by his other customers. The AIAG refers to this set of standards as the Inventory Visibility and Interoperability program (IV&I) [1]. And it has chosen to base the IV&I standards on existing work in manufacturing systems interoperability in the Open Applications Group [17].

The e-Kanban Process

One of the IV&I standards is directed to supporting the "e-Kanban" business process. The e-Kanban business process is a particular vendor-managed inventory scheme that is based on the Kanban system for managing materials flow in a manufacturing facility [2]. In the basic Kanban system, each time a part is consumed in a downstream process, a corresponding "Kanban ticket" is given to the upstream producer process for that part, and no producer process produces a part without a ticket authorizing that production. In practice, this process is buffered to some extent, but the idea is that all production is driven directly by documented active demand, from the final product all the way back to the materials sourcing. And the e-Kanban process is a means of implementing that demand-driven behavior at the materials sourcing end.

The e-Kanban process is based on an agreement between a manufacturer, as customer, and a supplier of some set of parts regularly used in the manufacturer's assemblies. Under the agreement, parts are shipped to one or more of the customer's facilities on the basis of actual consumption at the customer site. For each part covered by the agreement, a certain quantity of the part is always packaged as a unit for shipment. These shipment units are called *Kanbans*. As in the original Kanban process, a Kanban is shipped only when the customer sends an electronic "Kanban ticket" for that Kanban. The customer controls the flow of Kanban tickets according to his actual manufacturing demand for the part. A shipment consists of one or more Kanbans of one part intended for one customer facility. When the same supplier facility is supplying multiple small parts to the customer, the freight carrier may carry shipments of several different parts in one vehicle; at the other extreme, the Kanban may be an entire trailer or railroad car full of one part.

It is common for the supplier, or sometimes the customer, to dedicate a fixed set of industrial shipping containers to maintaining this part supply. At any given time, some of the containers are at the customer site being unloaded as needed to feed the assembly processes; some are at the supplier site being loaded with new parts as they are produced; and some are in transit between the supplier and customer sites. And when the customer empties a container, it is shipped back to the supplier to be refilled. For this reason, the process for supplying any one part is called a *Kanban Loop*. (Of course, in many cases, the physical shipping container is just part of the dunnage, and is discarded by the customer, but the term *Kanban loop* is still used.) So the business agreement, and the corresponding e-Kanban process covers one or more Kanban loops, one for each different part, and usually one for each distinct receiving facility.

This document details the specification for the business entities and properties that are involved in the e-Kanban business process, as defined by the IV&I program. It is these entities and properties that are the foundation for the information units that are exchanged in the standardized messages between the partner systems.

The IV&I specifications assume that the business agreements that create the e-Kanban process have already been put in place. (Another publication will document that process.) These agreements, and the necessary relationships with carriers and physical transportation, are referred to by "DocumentReferences" in the IV&I process and its ontology. The scope of this ontology is therefore strictly the entities and properties involved in the management of the materials flow. Because of the wide variance in the nature of the shipments in Kanban loops, many of the concepts specified in this ontology are somewhat general, and many of the concepts do not apply to all, or even most, e-Kanban processes.

This specification is divided into 8 sections, which relate to different aspects of the inventory visibility enterprise.

Ontologies and the presentation form

The term *ontology* is used here to mean a formal specification of the entities in some domain of interest, and the properties of those entities that are relevant to that interest [3], as indicated above. An ontology differs from an information model (such as an entity-attribute-relationship model or an object model) in one important way: The formal specification must be sufficiently well-defined that it can be used by a "computational reasoning system" to make valid inferences from known facts. This requires that the language used for the formal specification be defined in terms of a formal basis for logical reasoning.

The intended formal language for this specification is the Web Ontology Language (OWL) [4]. And the formal specification of the ontology in the OWL language appears as Annex A. The text of this specification, however, presents the ontology graphically, and documents each of the entities and properties in some detail. The graphical language used is "class diagrams" in the Unified Modeling Language (UML) [5].

This model uses five kinds of ontology elements:

- *Datatypes* represent classifications of pure information units that are represented directly in data repositories and messages. They are represented by UML "classes" with the "stereotypes":

    «atomic»   indicating an information unit with a simple computational representation.

    «enumeration»   indicating an information unit with a specified set of possible values «structure» indicating a conceptually atomic information unit that is composed of multiple simple computational elements.

A datatype may be said to be a *subtype* of another datatype, called a *supertype*, if all of its values are also values of the supertype. That is, the extension of the datatype is a subset of the extension of the supertype, and the interpretation of the datatype may be more specialized.

A datatype is said to be *abstract*, if all of the corresponding information units are further classified into subtypes.

- *Classes* represent classifications of business entities and other concepts that are not directly represented as data. They are represented by UML classes without stereotypes.

  A class is said to be *subsumed* by another class, called a *generalization*, if every individual in the class is also an instance of the generalization class and therefore has all of the modeled properties of that class as well. For each class, the relationship to a class that subsumes it is shown by a UML "subtype arrow" with a closed arrowhead.

  A class is said to be *abstract* if all of the individuals that are instances of the class have a narrower classification as well.

- *Datatype Properties* represent properties of an individual entity that can be represented directly by information units conveying the value of the property. These are represented by UML "attributes" inside the "class" box.

  Unless otherwise indicated, a datatype property has a *cardinality* of one, that is, every individual entity in the class has exactly one value for the property. If some instances of the class may not have the property at all, the property is shown with the suffix [0..1], i.e., at least 0 and at most 1.

  A few datatype properties are shown using the form for object properties (see below), where the range is a «structure» datatype.

- *Object Properties* represent relationships between the individual entities that are classified by the classes. Each object property is a relationship from instances of one class, called the *domain*, to instances of another class, called the *range*. These are represented by open-ended arrows (UML "directed associations") between the classes involved. Many object properties are said to have "inverses": a corresponding property of the range class that relates it to objects in the domain class and yields exactly the same pairings of individuals, but with the positions reversed. A property and its inverse are shown in the diagrams as a UML "association" line between the two class boxes with property names at each end and no arrowheads.

  Every object property has a *cardinality* specification, shown next to the property name. This indicates the number of individuals in the range that must be so related to each individual in the domain:

1 or 1..1	exactly one
0..1	at most one (minimum = 0, maximum = 1)
1..*	at least one (minimum = 1, maximum = unbounded)
* or 0..*	no constraint (minimum = 0, maximum = unbounded)

  Note: While UML allows a 1-to-many relationship to specify that it is ordered or unordered, the corresponding object properties are always unordered. In an ontology, an ordering is a property in its own right.

  A datatype property or object property can be described as a *Subproperty* of another property of the same kind. This means that every relationship that corresponds to the subproperty is also an instance of the other property, but it has additional characteristics. As a consequence, the range may be narrower, the cardinality may be more restricted, and one may have a useful inverse while the other does not.

- *Instances* represent individual entities in the domain of interest. Each instance has a name and a primary classification. They are not shown in UML diagrams.

Text Conventions

A term beginning with an upper-case letter (and possibly containing embedded upper-case letters) refers to a class or datatype defined in the ontology, e.g., "BusinessProcess". In a few cases, similar terms that do not begin with upper-case letters also appear in the descriptions of classes and properties, e.g., "business process". Such a term refers to a more general understanding of the concept, without the careful restrictions of the ontology definition.

A term beginning with a period refers to a defined property of the class being described. A term separated from a class name by a period refers to a defined property of that class.

Acknowledgement

The authors wish to thank the AIAG IV&I Business Process team – Patsy Snack, Steve Rudelic, Mary Kay Blantz, Irv Chmielewski, Paul Lewandowski, Manuel Porada, Shang-Tae Yee, and Mohammed Abidi – who provided the automotive supply-chain expertise for this specification, and reviewed and corrected it.

2 Basic Model Elements

This section defines the basic information unit concepts that correspond to computational datatypes. The concepts in this section are depicted in Figure 1 and described in detail below. The shaded datatypes in Figure 1 are considered fundamental data types.

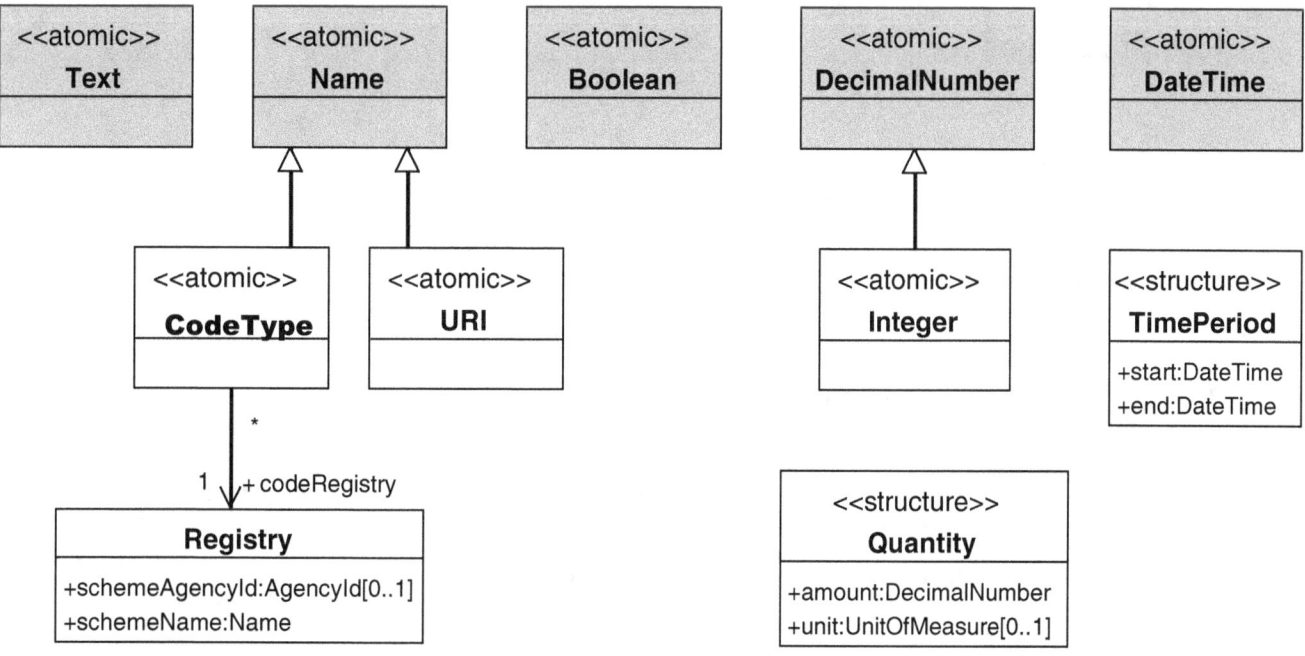

Figure 1   Basic Datatypes

2.1 Datatypes

2.1.1 Datatype: Boolean

Type: `atomic`

Definition: The data type for the truth values: true and false

2.1.1.1 Supertypes

none.

2.1.1.2 Elements

none.

2.1.1.3 Range   Roles

basic data type, not listed.

2.1.2 Datatype: CodeType

Type: `atomic`

Definition: A data element whose value is a "code" -- a string of characters whose meaning is specified in some reference standard or repository. In use, the reference standard/repository may be explicit or implicit.

Note: Technically, there is a relationship between a CodeType and some Registry – a document or database – that maintains the relationships between the code values and their meanings. This relationship is not modeled here, because it would require the CodeType to become a Class, so that it can have such ObjectProperties. And since CodeTypes are by definition types of data units, this seems to be the lesser of the evils.

Properties: abstract

2.1.2.1 Supertypes

Name

2.1.2.2 Elements

none.

2.1.2.3 Range   Roles

none.

2.1.3 Datatype: DateTime

Type: `atomic`

Definition: A data element representing a specific point in time, designated by a Date (usually in the Roman/Gregorian calendar) and possibly a Time of Day.

Note: The standard representation of this datatype is specified by ISO 8601.

2.1.3.1 Supertypes

none.

2.1.3.2 Elements

none.

2.1.3.3 Range   Roles

basic data type, not listed.

2.1.4 Datatype: DecimalNumber

Type: `atomic`

Definition: A number that may have a fractional part, usually expressed in decimal places.

2.1.4.1 Supertypes

none.

2.1.4.2 Elements

none.

2.1.4.3 Range   Roles

basic data type, not listed.

2.1.5 Datatype: Integer

Type: `atomic`

Definition: A data element representing a mathematical integer

2.1.5.1 Supertypes

DecimalNumber

2.1.5.2 Elements

none.

2.1.5.3 Range Roles

basic data type, not listed.

2.1.6 Datatype: Name

Type: `atomic`

Definition: A character string that is used to name something, and therefore can be compared for equal/unequal.

2.1.6.1 Supertypes

none.

2.1.6.2 Elements

none.

2.1.6.3 Range Roles

basic data type, not listed.

2.1.7 Datatype: Quantity

Type: `structure`

Definition: Conceptually a single information unit representing an amount of something. In practice it has two component information units: a number and a UnitOfMeasure.

2.1.7.1 Supertypes

none.

2.1.7.2 Elements

Element: amount Type: DecimalNumber

Definition: The numeric part of the Quantity, representing the number occurrences of the UnitOfMeasure that make up the quantity.

Cardinality: 1..1

Element: unit Type: UnitOfMeasure

Definition: The unit of measure for the Quantity.

Cardinality: 0..1

2.1.7.3 Range   Roles

basic data type, not listed.

2.1.8 Datatype: Text

Type: `atomic`

Definition: Text represents any set of character strings intended for consumption by a specific kind of agent, usually a human agent. Unlike a Name or Identifier, a Text is not intended to be compared for equal/unequal. It is intended to be processed by a specific kind of agent, and it may use a language and syntax that is appropriate to that kind of agent (and perhaps no other). Text to be presented to human agents uses an implicit or explicit natural language.

2.1.8.1 Supertypes

none.

2.1.8.2 Elements

none.

2.1.8.3 Range   Roles

basic data type, not listed.

2.1.9 Datatype: TimePeriod

Type: `structure`

Definition: A period of time that occurs at a specific (or repeating) point in time. (It is distinguished from a Duration, which is just a Quantity of time.) In the general case, a TimePeriod has a start time, and it may or may not have a given end time, because the termination of the TimePeriod is associated with some function and may not yet be established, even when the start time has passed.

2.1.9.1 Supertypes

none.

2.1.9.2 Elements

Element: start   **Type:** DateTime

Definition: The date/time at which the TimePeriod begins.

Cardinality: 1..1

Element: end   **Type:** DateTime

Definition: The date and time at which the TimePeriod (and the function/opportunity it delimits) ends. In some cases, this value is established at a later time than the start time, if ever. When the end date and time is not provided, it is considered to mean "some as yet unspecified future time".

Cardinality: 0..1

2.1.9.3 Range   Roles

basic data type, not listed.

2.1.10 Datatype: URI

Type: `atomic`

Definition: A Uniform Resource Identifier (URI), per IETF RFC 2396 (and other specific standards) that identifies an information "resource" that is currently available on, or accessible from some agency via, the World Wide Web.

2.1.10.1 Supertypes

Name

2.1.10.2 Elements

none.

2.1.10.3 Range Roles

none, used only in defining Email

2.1.11 Datatype: UnitOfMeasure

Type: `atomic`

Definition: A code designating a standard Unit of Measure or a derived unit of measure, as specified by ISO 31 and ISO 1000, or a common commercial unit of measure whose accuracy and relationship to SI units is well (often legally) defined.

The recommended code registry is UN/EDIFACT DE 6411.

2.1.11.1 Supertypes

CodeType

2.1.11.2 Elements

none.

2.1.11.3 Range Roles

none, used only in defining Quantity

2.2 Classes

none.

2.3 Instances

none.

3 Processes and Roles

This section introduces the fundamental concepts for business process models and their specializations for the IV&I e-Kanban business process. They are depicted in Figure 2 and Figure 3 and described in detail below.

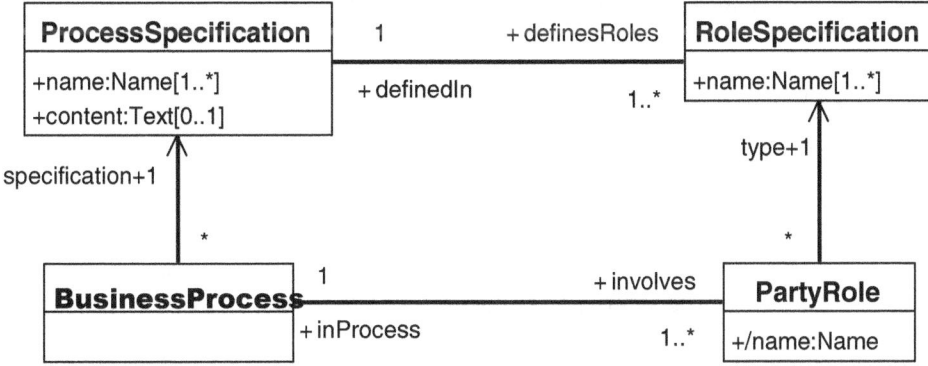

Figure 2   Business Processes

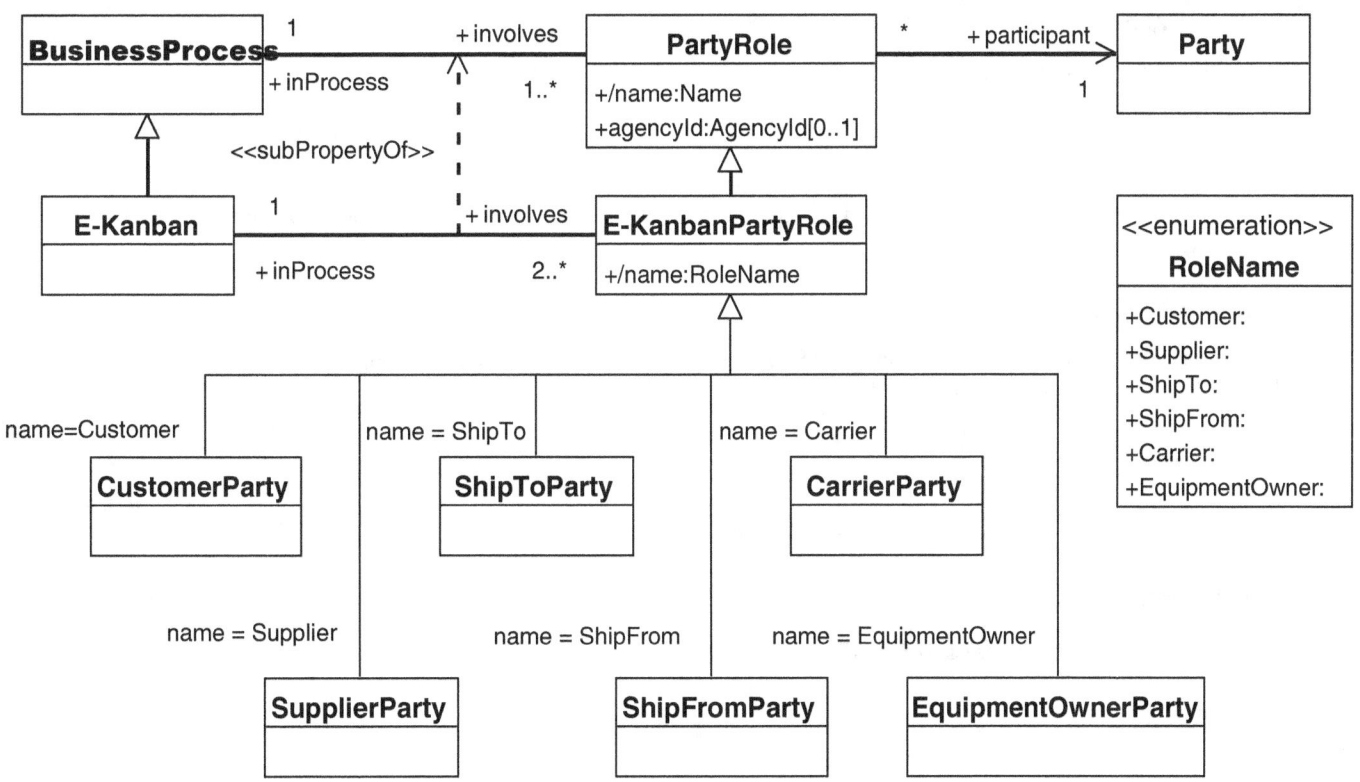

Figure 3   E-Kanban Process and Roles

3.1 Datatypes

3.1.1 Datatype: RoleName

Type: `enumeration`

Definition: A data unit that specifies the name of a role in an OAG business process. In the e-Kanban process the possible values are: Customer, Supplier, ShipFrom, ShipTo, Carrier, EquipmentOwnerParty

3.1.1.1 Supertypes

none.

3.1.1.2 Values

Value: Customer

Definition: Identifies the Party that buys/acquires goods or services in the e-Kanban process. See CustomerParty.

Value: Supplier

Definition: Identifies the Party that sells/provides goods or services in the e-Kanban process. See SupplierParty.

Value: Carrier

Definition: Identifies the Party responsible for movement of items between the Supplier and the Customer. See CarrierParty.

Value: ShipFrom

Definition: Identifies the Party, representing a Supplier facility, that functions as the Shipper in a KanbanLoop. See ShipFromParty.

Value: ShipTo

Definition: Identifies the Party, representing a Customer facility, that is the receiver of the Kanbans in a KanbanLoop See ShipToParty.

Value: EquipmentOwner

Definition: Identifies a Party that owns equipment (containers, trailers, etc.) that is used to hold goods in performing a Shipment, and possibly left on site, in a material replenishment process. See EquipmentOwnerParty.

3.1.1.3 Range Roles

From E-KanbanPartyRole as name

3.2 Classes

3.2.1 Class: BusinessProcess

Definition: A BusinessProcess is a specified behavior pattern for the conduct of some aspect of a business. It describes the activities a business and its agents perform in pursuing its objectives. For OAG purposes, a BusinessProcess of interest always involves interactions of multiple agents, and the "business" in question may be a "virtual enterprise" -- a process involving two autonomous business partners with some joint objective.

Properties: abstract

3.2.1.1 Generalizations

none.

3.2.1.2 Datatype Properties

none.

3.2.1.3 Object Properties

Property: specification   **Range: ProcessSpecification**

Definition: The specification that defines the principal activities and flows in the (joint) business process.

Cardinality: 1..1

Property: involves   **Range: PartyRole**

Inverse: PartyRole.inProcess

Definition: The actual PartyRoles, instance of the roles defined by the RoleSpecifications, that occur in this BusinessProcess instance, and by extension, the Parties that actually play them.

Cardinality: 1..*

3.2.1.4 Range Roles

From Message as occursIn

3.2.2 Class: CustomerParty

Definition: The party that buys/acquires goods or services in the e-Kanban business process, the Customer party:
- Consumes Inventory
- Establishes Kanban rules and values.
- Provides information on the Kanban status (consumed, authorized, etc.) and replenishment requirements
- Optionally, provides delivery receipt messages
- May monitor visibility tool for replenishment, in-transit, consumption, and alerts
- Collaborates on exception management with Supplier
- Performs planner function
- Performs receiving functions

3.2.2.1 Generalizations

E-KanbanPartyRole

3.2.2.2 Datatype Properties

none.

3.2.2.3 Object Properties

none

3.2.2.4 Range Roles

From: E-Kanban as customer

From: ShipmentSchedule as customer

An Ontology for the e-Kanban Business Process

3.2.3 Class: E-Kanban

Definition: A common, electronic signaling process for Kanban inventory replenishment schemes that can be enabled for both large and small trading partners in the automotive industry supply chain.

3.2.3.1 Generalizations

BusinessProcess

3.2.3.2 Datatype Properties

none.

3.2.3.3 Object Properties

Property: involves   Range: E-KanbanPartyRole

SubpropertyOf: BusinessProcess.involves

Inverse: E-KanbanPartyRole.inProcess

Definition: the refinement of BusinessProcess.involves that relates an e-Kanban business process to its PartyRoles.

Cardinality: 2..*.

Property: customer   Range: CustomerParty

SubpropertyOf: E-Kanban.involves

Definition: relationship between the e-Kanban process and the (sole) CustomerParty.

Note: An e-Kanban business process involves exactly one Customer.

Cardinality: 1..1

Property: supplier   Range: SupplierParty

SubpropertyOf: E-Kanban.involves

Definition: relationship between the e-Kanban process and the (sole) SupplierParty.

Note: An e-Kanban business process involves exactly one Supplier.

Cardinality: 1..1

Property: forLoops   Range: KanbanLoop

Inverse: KanbanLoop.supportingProcess (See Figure 15)

Definition: the KanbanLoops that are operated by this e-Kanban process. These KanbanLoops are distinguished by Item, and possibly by ShipTo and ShipFrom locations;

Cardinality: 1..*

3.2.3.4 Range Roles

none.

3.2.4 Class: E-KanbanPartyRole

Definition: Any of the PartyRoles that is defined for the e-Kanban business process. The .name of an E-KanbanPartyRole is restricted to one of: Customer, Supplier, ShipTo, ShipFrom, Carrier, EquipmentOwnerParty

Note: The CustomerParty and SupplierParty roles are documented in this section. The ShipToParty and ShipFromParty roles are documented in section 9. The CarrierParty and EquipmentOwnerPartyParty roles are documented in section 10.

3.2.4.1 Generalizations

PartyRole

3.2.4.2 Datatype Properties

Property: name **Range:** RoleName

SubpropertyOf: PartyRole.name

Definition: One of the 6 role names that refer to roles in the OAG/AIAG IV&I business process specification.

Cardinality: 1..1

3.2.4.3 Object Properties

Property: inProcess **Range:** E-Kanban

SubpropertyOf: PartyRole.inProcess

Inverse: E-Kanban.involves

Definition: the refinement of PartyRole.inProcess that relates an E-KanbanPartyRole to an e-Kanban business process.

Cardinality: 1..1

3.2.4.4 Range Roles

none.

3.2.5 Class: PartyRole

Definition: PartyRole is the BusinessActor concept as instantiated in an instance of an OAG-BusinessProcess. A PartyRole represents a Party playing a defined role (as modeled by a RoleSpecification) in a given BusinessProcess instance.

3.2.5.1 Generalizations

none.

3.2.5.2 Datatype Properties

Property: agencyId **Range:** AgencyId

Definition: A special code value from UN/CEFACT 3055 that designates a role in a business process.

Cardinality: 0..1

Property: name **Range:** Name

Definition: The name of the PartyRole, as given in the business process specification, e.g. Customer.

Properties: Derived from PartyRole.type and RoleSpecification.name

Cardinality: 1..1

3.2.5.3 Object Properties

Property: facility Range: Facility

Definition: The specific facility (Location) at which the Party conducts the business activities that fulfill this Role in this process. In general, there may be many, and they may or may not be identified. For the purposes of the e-Kanban process, certain roles specify single facilities as shipping and receiving points implicitly associated with the Party(Id).

Cardinality: 0..1

Property: participant Range: Party

Definition: The Party (organization or organizational unit) that plays this role in this process instance.

Cardinality: 1..1

Property: type Range: RoleSpecification

Definition: The nominal Role (in the defined business process) that this PartyRole instantiates with an actual participant Party.

Cardinality: 1..1

Property: contacts Range: Contact

Inverse: Contact.forPartyInRole

Definition: Persons or organizational units designated as representatives of the participant Party in this role in this process instance.

Cardinality: 1..*

Property: inProcess Range: BusinessProcess

Inverse: BusinessProcess.involves

Definition: The BusinessProcess instance in which this PartyRole actually exists.

Cardinality: 1..1

3.2.5.4 Range Roles

none.

3.2.6 Class: ProcessSpecification

Definition: A specification for a business process that defines the actors (Roles), the expected behaviors, the sequence of actions, the decision points, etc. It can be specified in some diagrammatic language, or some formal language, or natural language, or (usually) some combination of the above.

3.2.6.1 Generalizations

none.

3.2.6.2 Datatype Properties

Property: content Range: Text

Definition: The details of the specification for the process, possibly given in some formal language.

Cardinality: 0..1

Property: name **Range:** Name

Definition: The name of the specification, and by extension, the name of the reference business process.

Cardinality: 1..*

3.2.6.3 Object Properties

Property: definesRoles **Range:** RoleSpecification

Inverse: RoleSpecification.definedIn.

Definition: The nominal players in a process that conforms to the ProcessSpecification. The specification gives these roles names and specifies their responsibilities in the process.

Cardinality: 1..*

3.2.6.4 Range Roles

From: BusinessProcess **as specification**

3.2.7 Class: RoleSpecification

Definition: That part of a ProcessSpecification that defines one Role in the process and gives it a name.

3.2.7.1 Generalizations

none.

3.2.7.2 Datatype Properties

Property: name **Range:** Name

Definition: The name of the Role that is defined.

Cardinality: 1..*

3.2.7.3 Object Properties

Property: definedIn **Range:** ProcessSpecification

Inverse: ProcessSpecification.definesRoles.

Definition: The specification that includes the Role being defined.

Cardinality: 1..1

3.2.7.4 Range Roles

From: PartyRole **as type**

3.2.8 Class: SupplierParty

Definition: the Party information for the Party that sells/provides goods or services in the e-Kanban business process. The SupplierParty.:
- Performs fulfilment function
- Performs shipping function
- May monitor visibility tool for replenishment, in-transit, consumption, and alerts
- Collaborates on exception management with Customer.

3.2.8.1 Generalizations

E-KanbanPartyRole

3.2.8.2 Datatype Properties

none.

3.2.8.3 Object Properties

none.

3.2.8.4 Range Roles

From: E-Kanban as supplier

From: ShipmentSchedule as supplier

3.3 Instances

none.

4 Identifiers

This section defines the OAG Identifier concept – a designation for a specific thing, such as an organization, a document, or a manufactured item, that is maintained in a public or private Registry. In an e-business process, the same thing can be identified by many different designations, according to the agreements of the Parties involved. The concepts in this section are depicted in Figure 4 and described in detail below.

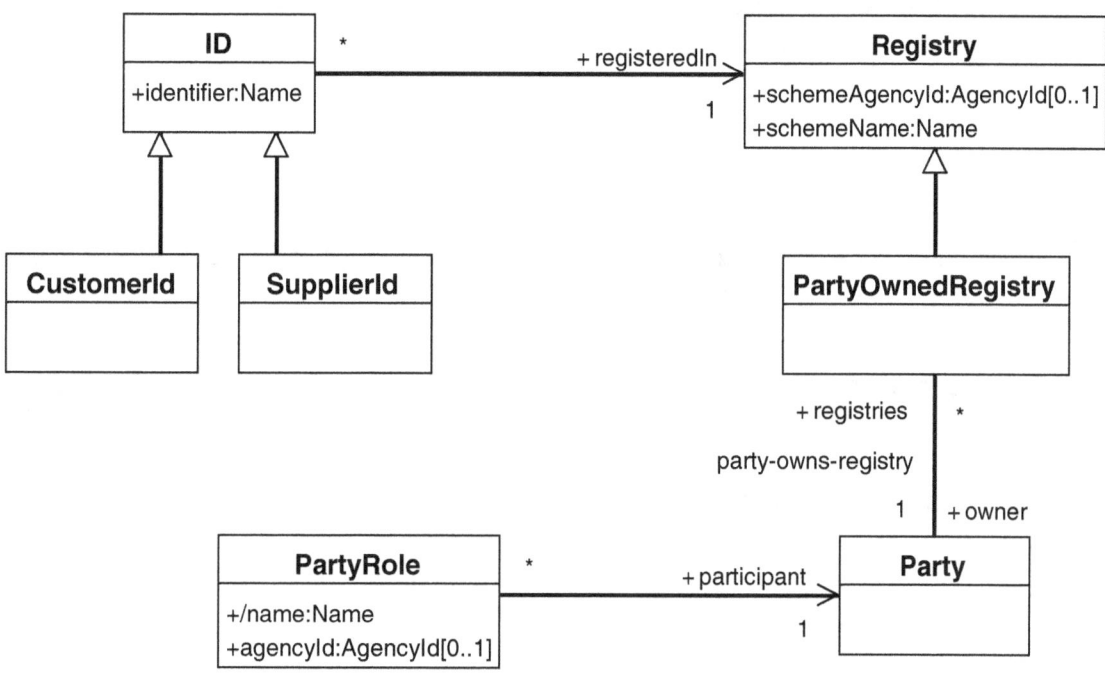

Figure 4   Registered Identifiers

4.1 Datatypes

4.1.1 Datatype: AgencyId

Type: atomic

Definition: A code from UN/CEFACT 3055 [6], designating an agency or registry that owns a code list or assigns identifiers to organizations. Most values from the UN/CEFACT code list designate specific international registries of businesses or codes. Certain special values designate roles in a business process.

4.1.1.1 Supertypes

CodeType.

4.1.1.2 Values

Most values identify specific registries. In particular, the e-Kanban business process uses the Dun & Bradstreet Registry of Businesses. Two special codes are used to identify registries owned by parties to the business process.

Value 16

Definition: identifies the Dun & Bradstreet Registry of Businesses

Value 91

Definition: A special value that identifies the SupplierParty in the current business process.

Value 92

Definition: A special value that identifies the CustomerParty in the current business process

4.1.1.3 Range Roles

From Registry as schemeAgencyId

From PartyRole as agencyId

4.2 Classes

4.2.1 Class: CustomerId

Definition: An identifier (ID) for a Party, Document, etc., that was assigned by the CustomerParty to the OAG BusinessProcess. That is, an ID that is registeredIn a PartyOwnedRegistry owned by the Party that is the CustomerParty to the BusinessProcess.

4.2.1.1 Generalizations

ID

4.2.1.2 Datatype Properties

none.

4.2.1.3 Object Properties

none.

4.2.1.4 Range Roles

none.

4.2.2 Class: ID

Definition: A unique identifier for a business object/entity, together with the identification of the registry or other source of assignment that makes the identifier unique.

Properties: abstract.

4.2.2.1 Generalizations

none.

4.2.2.2 Datatype Properties

Property: identifier   **Type: Name**

Definition: the actual identifier value, the name for whatever the ID identifies

Cardinality: 1..1

4.2.2.3 Object Properties

Property: registeredIn   **Range:** Registry

Definition: the Registry that assigns a referent to the .identifier value.

Cardinality: 1..1

4.2.2.4 Range Roles

none. (All references are to subclasses of ID.)

4.2.3 Class: PartyOwnedRegistry

Definition: a Registry of identifiers that are assigned by a business organization, such as an Approved Supplier registry, or a registry of product model numbers or serial numbers.

Note: A PartyOwnedRegistry is not usually considered to have a schemeAgencyId. In e-business transactions, the PartyOwnedRegistry is usually identified only by the PartyRole played by the owner Party in that transaction. The agencyId for the PartyRole identifies the PartyRole and refers to the participant Party by extension. The schemeName may be implicit in the transaction, according to the type or role of the thing being identified.

4.2.3.1 Supertypes

Registry

4.2.3.2 Datatype Properties

none.

4.2.3.3 Object Properties

Property: owner   **Range:** Party

Definition: the Party that maintains the Registry

Cardinality: 1..1

4.2.3.4 Other Roles

none.

4.2.4 Class: Registry

Definition: A repository of codes or identifiers, together with their meaning, and possibly effectivity and provenance information. The agency that controls the registry assigns the code or identifier for that meaning, and guarantees that the code or identifier is unique across the registry.

4.2.4.1 Generalizations

none.

4.2.4.2 Datatype Properties

Property: schemeAgencyId   **Range:** AgencyId

Definition: A code (from UN/CEFACT 3055 [6]) that designates the agency or registry.

Note: PartyOwnedRegistries do not have .schemeAgencyIds. The special values of AgencyId designate PartyRoles, which refer indirectly to the Party playing that role in a given BusinessProcess instance, and thus to some corresponding registry maintained by that Party.

Cardinality: 0..1

Property: schemeName   Range: Name

Definition: The common name of the identification scheme, agency or registry.

Cardinality: 0..1

4.2.4.3 Object   Properties

none.

4.2.4.4 Range   Roles

From ID as registeredIn

4.2.5 Class: SupplierId

Definition: An Identifier for a business entity that is assigned by the SupplierParty in the business process. That is, an ID that is registeredIn a PartyOwnedRegistry owned by the Party that is the SupplierParty to the BusinessProcess.

4.2.5.1 Generalizations

ID

4.2.5.2 Datatype   Properties

none.

4.2.5.3 Object   Properties

none.

4.2.5.4 Range   Roles

none.

4.3 Instances

4.3.1 DUNS

Type: Registry

Definition: represents the Dun & Bradstreet Registry of Businesses

Facts

schemeName(DUNS) = "DUNS"

schemeAgencyId(DUNS) = 16

5 Parties

This section introduces the concept Party, meaning a partner to the business process, and its related concepts. These are depicted in Figure 5 and described in detail below.

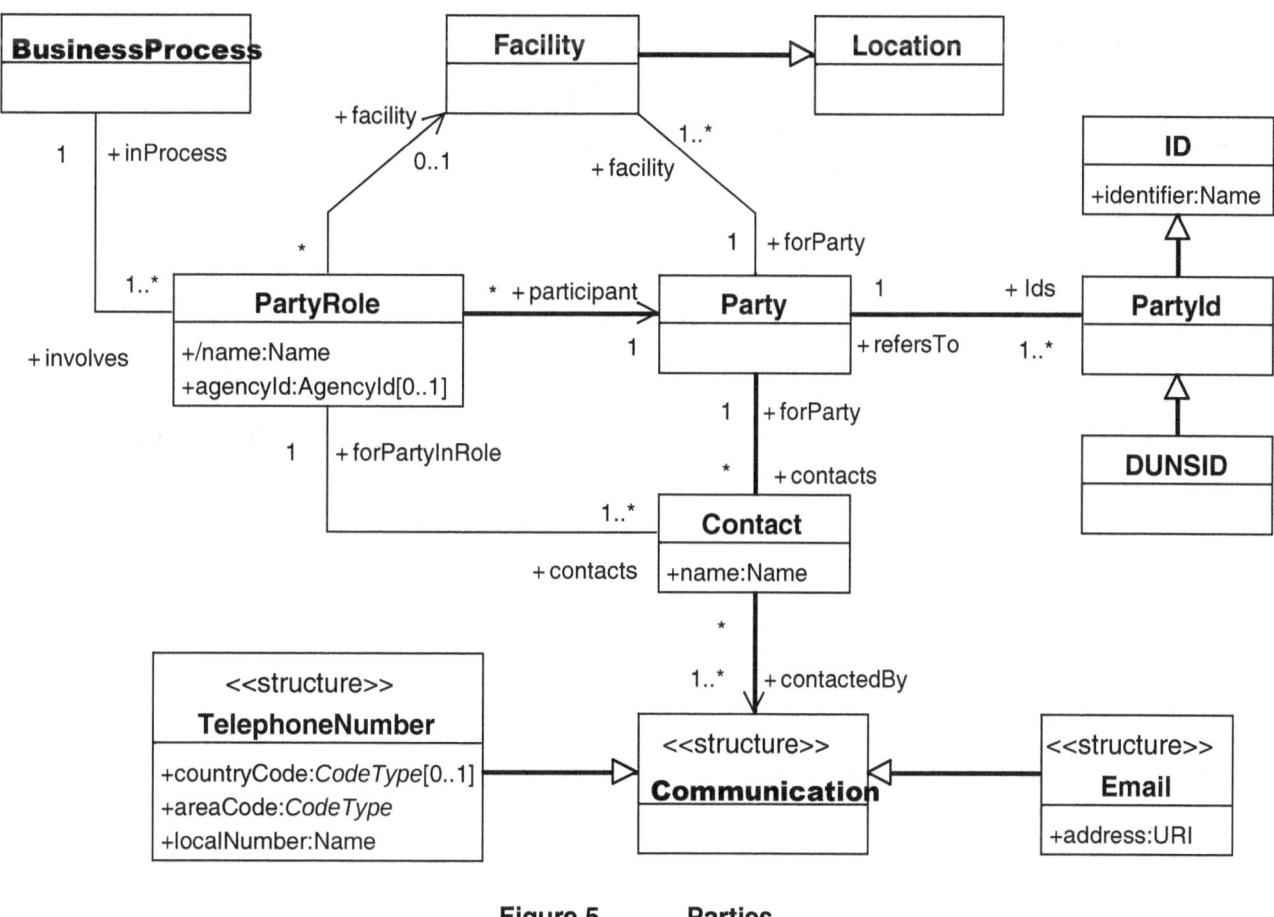

Figure 5   Parties

5.1 Datatypes

5.1.1 Datatype: Communication

Type: `structure`

Definition: A data element or elements that identify a communications "channel" -- the necessary codes for using a means of making contact with a person or organization. Examples: postal address, email address, telephone number, etc.

Properties: Abstract.

5.1.1.1 Supertypes

none.

5.1.1.2 Elements

none.

5.1.1.3 Range Roles

From: Contact as contactedBy

5.1.2 Datatype: Email

Type: `structure`

Definition: An email address -- the means of access to a Person or Organization by electronic mail.

5.1.2.1 Supertypes

Communication

5.1.2.2 Elements

Element: address   **Type: URI**

Definition: An email address corresponding to IETF 2368 [7], but in the URI form, which begins: "mailto:"

Cardinality: 1..1

5.1.2.3 Range Roles

none.

5.1.3 Datatype: TelephoneNumber

Type: `structure`

Definition: a telephone number, complete with all the necessary dialing information needed for the business use

5.1.3.1 Supertypes

Communication

5.1.3.2 Elements

Element: areaCode   **Type: CodeType**

Definition: An "area code" or "region code" as defined by the national telephone system

Cardinality: 1..1

Element: countryCode   **Type: CodeType**

Definition: an ITU-T country code, designating the (national) telephone system being called

Cardinality: 0..1

Element: localNumber   **Type: Name**

Definition: the local telephone number within the area

Cardinality: 1..1

5.1.3.3 Range Roles

none.

5.2 Classes

5.2.1 Class: Contact

Definition: A person or organizational unit acting for a Party in a given PartyRole. Contact is a role defined only for a given PartyRole. That is, if the Party is engaged in the same process with two different business partners and uses the same Person as the "contact" for both, that is two different "Contacts" using the same Person (who is not modeled here).

5.2.1.1 Generalizations

none.

5.2.1.2 Datatype Properties

Property: name Range: Name

Definition: The name of the Person or Group (organizational unit) that represents the Party in this role.

Cardinality: 1..1

Property: contactedBy Range: Communication

Definition: The means of communication with the Contact -- telephone, email, etc. There may be several. This model does not distinguish different roles of the same mechanism, e.g. telephone numbers for office, secretary, cell, FAX.

Cardinality: 1..*

5.2.1.3 Object Properties

Property: forPartyInRole Range: PartyRole

Inverse: PartyRole.contacts

Definition: The role in a given BusinessProcess instance for which this Person is a Contact. This association really defines the Contact role.

Cardinality: 1..1

Property: forParty Range: Party

Inverse: Party.contacts

Definition: The Party represented by the Contact in this PartyRole

Cardinality: 1..1

5.2.1.4 Range Roles

none.

5.2.2 Class: DUNSID

Definition: A PartyId representing the DUNS number -- the identification for the business or division that was assigned by the Dun&Bradstreet registry of businesses. That is, a PartyId that is registeredIn the DUNS Registry.

5.2.2.1 Generalizations

PartyId

5.2.2.2 Datatype Properties

none.

5.2.2.3 Object Properties

none.

5.2.2.4 Range Roles

none.

5.2.3 Class: Facility

Definition: A Location at which a particular Party does business.

Note: For the AIAG IV&I project, it is assumed that the location information associated with a Facility, e.g., its address, is implicit in the identification of the Party playing the PartyRole in which that Facility is involved.

5.2.3.1 Generalizations

Location

5.2.3.2 Datatype Properties

none.

5.2.3.3 Object Properties

Property: forParty   **Range:** Party

Inverse: Party.facility

Definition: The Party who does business at this Location.

Cardinality: 1..1

5.2.3.4 Range Roles

From: PartyRole **as facility**

5.2.4 Class: Party

Definition: A Person or Organization, as that individual may be involved directly in a business process. For this purpose, a Party has one or more unique PartyIds and one or more business locations (facilities).

5.2.4.1 Generalizations

none.

5.2.4.2 Datatype Properties

none.

5.2.4.3 Object Properties

Property: facility   **Range:** Facility

Inverse: Facility forParty

Definition: A location at which the Party does business.

Cardinality: 1..*

Property: usesSystem   **Range:** System

Inverse: System agentFor (see Figure 16)

Definition: The relationship between the Party and the software Systems it uses in the e-business processes. In general, this is the inverse of the interesting relationship.

Cardinality: 0..*

Property: contacts   **Range:** Contact

Inverse: Contact forParty

Definition: Any Person or organizational unit that serves as a Contact for the Party in some BusinessProcess.

Cardinality: 0..*

Property: Ids   **Range:** PartyId

Inverse: PartyId refersTo

Definition: The unique identifiers for the Party. This may include identifiers, such as Supplier id or Customer id, that are assigned by the process participants, in addition to formal identifiers, like tax-id and DUNS number.

Cardinality: 1..*

Property: registries   **Range:** PartyOwnedRegistry

Inverse: PartyOwnedRegistry owner

Definition: Identifier registries maintained by the Party, such as a Customer registry or an Approved Supplier registry

Cardinality: 0..*

5.2.4.4 Range Roles

From: PartyRole **as participant**

From: Message **as sender**

From: Message **as receiver**

5.2.5 Class: PartyId

Definition: A registered identifier for a Party (see ID). The most commonly used PartyId for a business unit is the DUNSID.

5.2.5.1 Generalizations

ID

5.2.5.2 Datatype Properties

none.

5.2.5.3 Object Properties

Property: refersTo   **Range:** Party

Inverse: Party_ids

Definition: The specific organization, or organizational unit, to which this Identifier refers.

Cardinality: 1..1

5.2.5.4 Range Roles

none.

5.3 Instances

none.

An Ontology for the e-Kanban Business Process   NISTIR 7404

6 Documents

This section introduces the basic Document concept and related concepts of documentation. These concepts are depicted in Figure 6 and described in detail below.

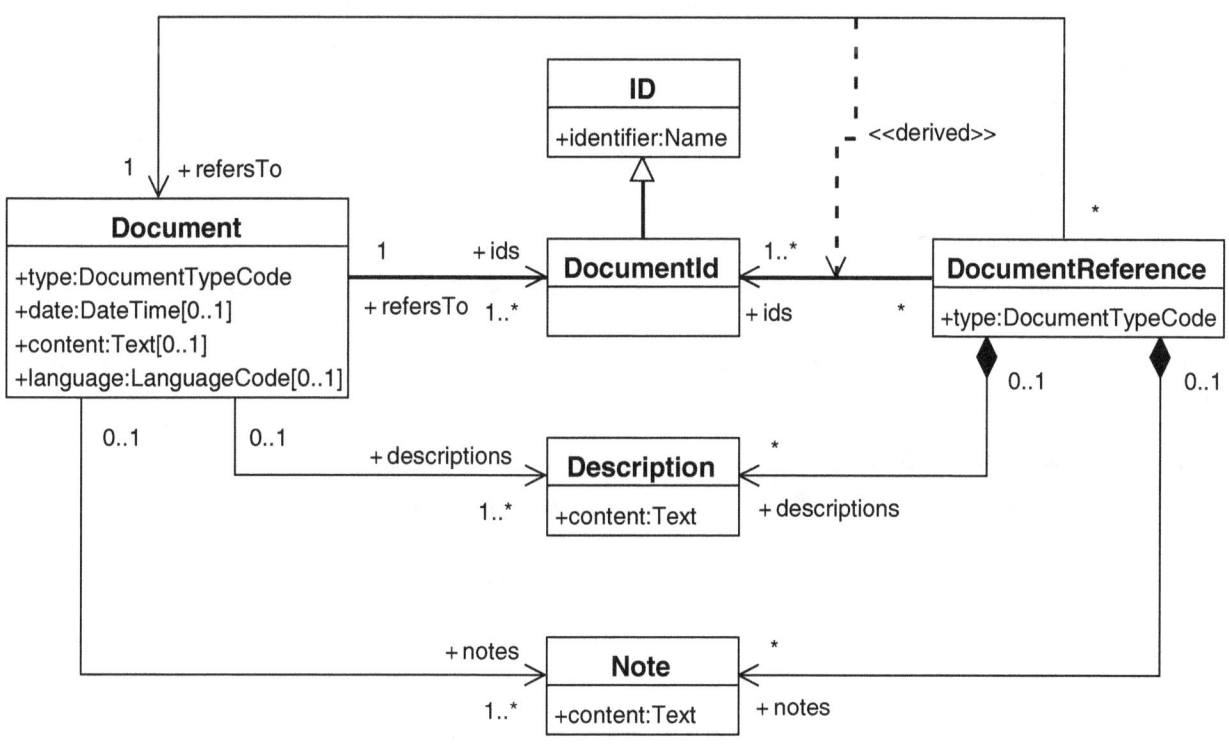

Figure 6   Documents

6.1 Datatypes

6.1.1 Datatype: DocumentTypeCode

Type: `atomic`

Definition: A CodeType that is used to identify the nature of a Document. The values are defined by UN/CEFACT 1001 [8].

6.1.1.1 Supertypes

CodeType

6.1.1.2 Values

Value: 105 Purchase order

Definition: Document/message issued within an enterprise to initiate the purchase of articles, materials or services required for the production or manufacture of goods to be offered for sale or otherwise supplied to customers.

Value: 173 Authorization to plan and ship orders

Definition: Document or message that authorizes receiver to plan and ship orders based on information in this message.

Value: 315 Contract

Definition: (1296) Document or message evidencing an agreement between the seller and the buyer for the supply of goods or services.

Value: 236 Delivery forecast

Definition: Document or message that informs the receiver of the anticipated product/delivery requirements of the sender over some period of time.

Value: 241 Delivery schedule

Definition: Document that defines the contractual, requested or anticipated shipments (quantities, dates/times) of one or more items under a purchase agreement.

Note: Note: For the e-Kanban process, the Document content should conform to ANSI X12.3 (EDI) [9] Message type 830 (Delivery Schedule)

Value: 351 Shipment advice ("Despatch advice")

Definition: Document or message by which the seller or consignor informs the consignee or receiver about a planned or initiated shipment of goods.

Note: Note: For the e-Kanban process, the Document content should be a ShipmentNotification (see **Error! Reference source not found.**)

Value: 705 Bill of lading

Definition: Negotiable document which evidences a contract of carriage by sea and the taking over or loading of goods by carrier, and by which carrier undertakes to deliver goods against surrender of the document. A provision in the document that goods are to be delivered to the order of a named person, or to order, or to bearer, constitutes such an undertaking.

Value: 786 Freight manifest

Definition: Document that describes the content of a carrier equipment item (such as a trailer, rail car, or ship), including product type, source, immediate destination, final destination, packaging and labels, current ownership and liability, and references to contractual and other authorizations and licenses. A Freight manifest contains the same information as a cargo manifest, and additional details on freight amounts, charges, etc.

6.1.1.3 Range   Roles

From Document **as type**

From DocumentReference **as type**

6.1.2 Datatype: LanguageCode

Type: atomic

Definition: A code (from ISO 639) that identifies the natural language of some text

6.1.2.1 Supertypes

CodeType

6.1.2.2 Elements

none.

6.1.2.3 Range Roles

From Document as language

6.2 Classes

6.2.1 Class: Description

Definition: A body of free text attached to an Item, Document, etc., for the purpose of conveying its nature, content, purpose, etc. to a human.

6.2.1.1 Generalizations

none.

6.2.1.2 Datatype Properties

Property: content   Range: Text

Definition: The body of text that provides the descriptive information

Cardinality: 1..1

6.2.1.3 Object Properties

none.

6.2.1.4 Range Roles

From: DocumentReference as descriptions

From: Document as descriptions

From: Item as descriptions

6.2.2 Class: Document

Definition: Any structured or unstructured collection of information that has identity. A Document has one or more identifiers that distinguish it from other documents. A document has a type, which indicates what kind of information it contains, and possibly what structure it has.

6.2.2.1 Generalizations

none.

6.2.2.2 Datatype Properties

Property: content   Range: Text

Definition: Represents the content of a Document that is unstructured text. This attribute is not present in a document that has a modeled structured form, such as a BusinessObjectDocument

Cardinality: 0..1

Property: date Range: DateTime

Definition: A non-specific "date of availability" associated with a Document. For documents representing business agreements, contractual or legal commitments and affadavits, this is the date of execution of the document. For "documents" constructed explicitly for messages, this is the "as of" date/time for the information contained in the document. For documents of other kinds, the date is intended to distinguish versions, and is typically a publication date, copyright date, or date of issuance.

Cardinality: 0..1

Property: type Range: DocumentTypeCode

Definition: the reference classification for the Document, which typically includes its nature and its structure.

Cardinality: 1..1

Property: language Range: LanguageCode

Definition: Specifies the (primary) natural language for the textual elements of the Document.

Cardinality: 0..1

6.2.2.3 Object Properties

Property: notes Range: Note

Definition: the relationship of the Document to Notes that relate to its status or usage.

Cardinality: 0..*

Property: descriptions Range: Description

Definition: the relationship between the Document and a text description of the document type, title, authors, etc. or to summaries or abstracts of the Document.

Cardinality: 0..*

Property: ids Range: DocumentId

Definition: the Identifiers that uniquely identify the Document.

Note: Documents have distinct IDs given to them by the parties involved, and possibly by governmental agencies with which they are filed. In some cases, different "versions" of a Document are in use and the "version" is encoded in the identifiers; in other cases the "version" is captured as the .date of the Document.

Cardinality: 1..*

6.2.2.4 Range Roles

From: DocumentReference **as refersTo**

6.2.3 Class: DocumentId

Definition: An ID used as an identifier for a Document.

6.2.3.1 Generalizations

ID

6.2.3.2 Datatype Properties

none.

6.2.3.3 Object Properties

none.

6.2.3.4 Range Roles

From: DocumentReference **as ids**

From: Document **as ids**

6.2.4 Class: DocumentReference

Definition: An information structure used to refer to one Document in a particular use of that document. It should always include the type and at least one identifier (DocumentId) for the document in question. A DocumentReference may also include some Description of the document (for the benefit of human readers), and it may include some Notes as to the use of the referenced document in the context of the reference. A DocumentReference always occurs physically in some other Document or Message. A DocumentReference is a physical body of text that is usually separate from the referenced Document. In some cases, a DocumentReference may be used when the Document to which it refers is not readily available, or not required to be readily available, such as a reference to a standard or a body of public law.

Note: No two DocumentReferences are considered to be the same individual, even when they are identical in content. A DocumentReference is a part of an individual information structure and is distinguished in part by the structure that includes it.

6.2.4.1 Generalizations

none.

6.2.4.2 Datatype Properties

Property: type   **Range:** DocumentTypeCode

Definition: the reference classification for the Document, which typically includes its nature and its structure. In many cases, a DocumentId may only be unique within the DocumentType

Cardinality: 1..1

6.2.4.3 Object Properties

Property: refersTo   **Range:** Document

Definition: the relationship between a DocumentReference and the unique Document to which it refers. This is the conceptual relationship, but its implementation uses DocumentReference.ids and DocumentId.refersTo Document.

Cardinality: 1..1

Property: notes   **Range:** Note

Definition: Notes that occur in or with the DocumentReference that relate to the status or usage of the Document in this context.

Cardinality: 0..*

Property: descriptions   **Range:** Description

Definition: text description of the Document: type, title, authors, etc. or summaries or abstracts of the Document, that are included in/with the DocumentReference. This information permits a human agent to determine, among a collection of DocumentReferences, which Reference is to the Document he is looking for, since the DocumentReference proper may be just an identifying number.

Cardinality: 0..*

Property: ids   **Range:** DocumentId

Definition: the document Identifiers contained in the DocumentReference. At least one must be present, and all of the Identifiers present refer to the same Document.

Cardinality: 1..*

6.2.4.4 Range   Roles

From: ShipmentSchedule **as references**

From: ScheduleLine **as references**

From: Shipment **as basisDocuments**

From: ShipmentUnit **as attachedDocuments**

6.2.5 Class: Note

Definition: A unit of free-form text attached to some business entity to add information for human consumption, usually about a particular usage of the entity.

6.2.5.1 Generalizations

none.

6.2.5.2 Datatype   Properties

Property: content

Type: Text

Definition: the text of the Note

Cardinality: 1..1

6.2.5.3 Object   Properties

Property: toItem   **Range:** Item

Inverse: Item.notes

Definition: the Item(s) to which the Note applies.

Note: The same Note can be explicitly applied to multiple Items and a change to the Note affects all of them.

Cardinality: 0..*

Property: toLine   **Range:** ScheduleLine

Inverse: ScheduleLine.notes

Definition: the ScheduleLine(s) to which the Note applies.

Note: The same Note can be explicitly applied to multiple ScheduleLines and a change to the Note affects all of them.

Cardinality: 0..*

Property: toPackaging   **Range:** StandardPack

Inverse: StandardPack.notes

Definition: the StandardPack specifications to which the Note applies.

Note: The same Note can be explicitly applied to multiple StandardPack specifications and a change to the Note affects all of them.

Cardinality: 0..*

6.2.5.4 Range Roles

From: LocationReference **as notes**

From: StandardPack **as notes**

From: Package **as notes**

From: DocumentReference **as notes**

From: Document **as notes**

6.3 Instances

none.

7 Locations

This section introduces the Location concept, as it is used in the e-Kanban business process. The related concepts are depicted in Figure 7 and described in detail below.

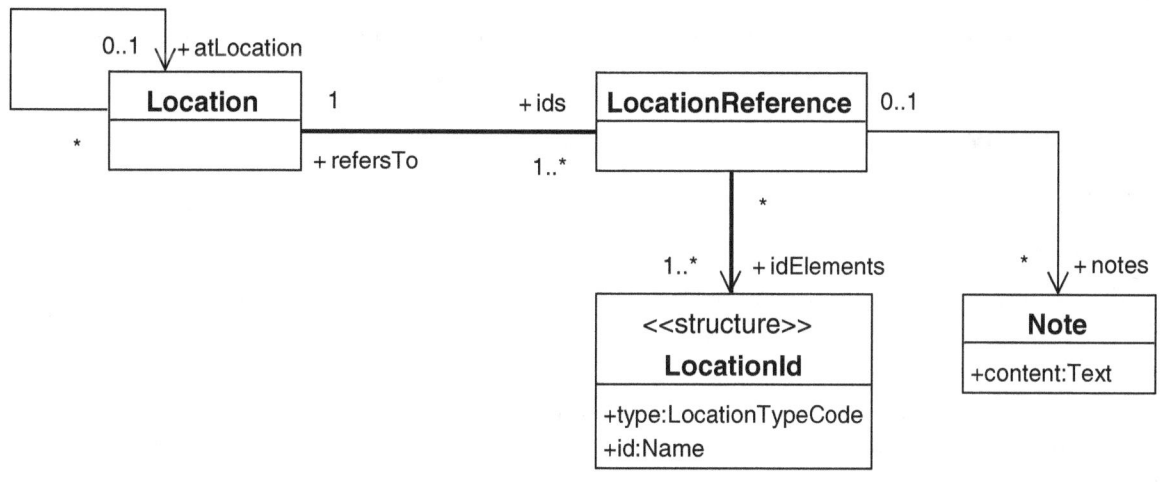

Figure 7   Locations

7.1 Datatypes

7.1.1 Datatype: LocationId

Type: structure

Definition: A LocationId is an identifier, or part of an identifier, for a Location. LocationId elements include GPS (latitude/longitude) codes, freight terminal codes, postal codes, country, city, state, street address, plant number, area, loading dock, etc. In general, every LocationId identifies a Location, but possibly only in the context of another (broader) Location.

7.1.1.1 Supertypes

none.

7.1.1.2 Elements

Element: id   **Type: Name**

Definition: The actual value of the location code.

Cardinality: 1..1

Element: type   **Type: LocationTypeCode**

Definition: A code from UN/CEFACT 3227 [10] that identifies the nature of the LocationCode element: FreightTerminal, GPS, PostalAddress, etc.

Cardinality: 1..1

7.1.1.3 Range Roles

From: LocationReference **as idElements**

7.1.2 Datatype: LocationTypeCode

Type: `atomic`

Definition: A code from UN/CEFACT 3227 [10] that identifies the nature of a LocationCode element, e.g. FreightTerminal, GPS, PostalAddressLine, etc.,

7.1.2.1 Supertypes

CodeType

7.1.2.2 Values

The following code values from UN/CEFACT 3227 are commonly used in the e-Kanban business process:

Value: 11

Definition: Place/port of discharge, e.g. Freight terminal: Airport, Seaport, Rail Terminal, etc. Corresponding LocationCode.id values are taken from UN/CEFACT Recommendation 16 (LOCODE) [11], e.g. "LAX".

Value: 141

Definition: loading dock

Value: 159

Definition: user-defined site location, such as a line-end or hopper.

7.1.2.3 Object Properties

none.

7.1.2.4 Range Roles

none.

7.2 Classes

7.2.1 Class: Location

Definition: A Location is a place of interest. For most purposes, a Location has one or more complex identifiers, and any other properties of interest are implied by the use of the Location.

7.2.1.1 Generalizations

none.

7.2.1.2 Datatype Properties

none.

7.2.1.3 Object Properties

Property: atLocation   **Range: Location**

Definition: A more general Location, to which this Location is subordinate and more precise.

Cardinality: 0..1

Property: ids   **Range: LocationReference**

Inverse: LocationReference refersTo

Definition: the relationship between the location and one LocationReference that identifies it. A Location may have more than one such identifier.

Cardinality: 1..*

7.2.1.4 Range Roles

From: Location as atLocation

From: Shipment as receivingPoint

7.2.2 Class: LocationReference

Definition: An identifier for a Location, to some degree of accuracy. A LocationReference consists of a set of LocationId elements that are intrinsically ordered conceptually: one LocationId specifies the most general Location and other LocationIds convey successively further refinements. In some cases, such as a FreightTerminal identifier or a GPS id, there is only one LocationId.

7.2.2.1 Generalizations

none.

7.2.2.2 Datatype Properties

Property: idElements   **Range: LocationId**

Definition: the relationship between the LocationReference and the LocationId elements that make up the reference.

Cardinality: 1..*

7.2.2.3 Object Properties

Property: notes   **Range: Note**

Definition: Notes attached to the LocationReference, that is, Notes that pertain to a particular reference to a Location, e.g. for access times or regulations, contacts, etc.

Cardinality: 0..*

Property: refersTo   **Range: Location**

Inverse: Location ids

Definition: The Location to which the LocationReference refers. The granularity of the Location depends on the details of the LocationReference.

Cardinality: 1..1

7.2.2.4 Range Roles
none.

7.3 Instances
none.

8 Items

This section introduces the concept Item, that is, a material or manufactured part or product. For the e-Kanban process, an Item is little more than its catalog entry, plus packaging concepts. The related concepts are depicted in Figure 8 and described in detail below.

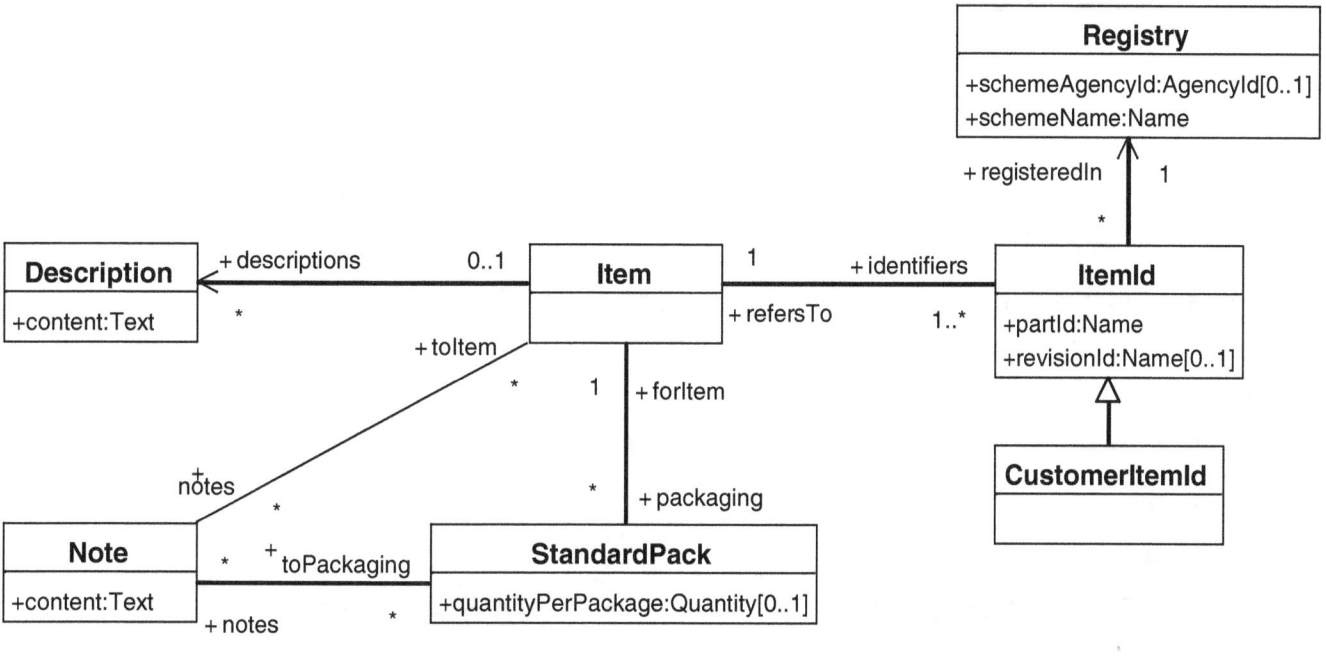

Figure 8   Items

8.1 Datatypes

none.

8.2 Classes

8.2.1 Class: CustomerItemId

Definition: An ItemId that was assigned to the Item by the CustomerParty to the OAG BusinessProcess. That is, an ItemId that is registeredIn a PartyOwnedRegistry owned by the Party that is the CustomerParty to the BusinessProcess.

8.2.1.1 Generalizations

ItemId

8.2.1.2 Datatype   Properties

none.

8.2.1.3 Object   Properties

none.

8.2.1.4 Range Roles
none.

8.2.2 Class: Item

Definition: Any inventory item -- a product, a catalog product, a part, a tool, a tooling part, a consumable item, etc.

8.2.2.1 Generalizations
none.

8.2.2.2 Datatype Properties
none.

8.2.2.3 Object Properties

Property: packaging   **Range: StandardPack**

Inverse: StandardPack.forItem

Definition: the StandardPack specifications for packaging the Item.

Cardinality: 0..*

Property: notes   **Range: Note**

Inverse: Note.toItem

Definition: the relationship between an Item and Notes that describe its handling, status, usage, etc.

Cardinality: 0..*

Property: descriptions   **Range: Description**

Definition: the relationship between an Item and its (text) Description(s)

Cardinality: 0..*

Property: identifiers   **Range: ItemId**

Inverse: ItemId.refersTo

Definition: the relationship between an Item and the ItemIds that refer to it.

Cardinality: 1..*

8.2.2.4 Range Roles

From: ScheduleLine as shipmentItem

From: KanbanLoop as suppliedItem

From: Shipment as shipmentItem

8.2.3 Class: ItemId

Definition: A unique identifier for the Item. An Item may have multiple identifiers, e.g., the part number assigned by the manufacturer and the part number assigned by the customer.

8.2.3.1 Generalizations

none.

8.2.3.2 Datatype Properties

Property: partId   **Range: Name**

Definition: the principal identifier for the Item, as distinct from a related revision/version id.

Cardinality: 1..1

Property: revisionId   **Range: Name**

Definition: Also called Engineering Change Level. This is an optional subordinate identifier used to distinguish versions of a Part, that are nominally compatible in "form, fit and function". A revision, or engineering change level, often represents minor differences in design, material or manufacture that may or may not affect suitability of the part for particular uses.

Cardinality: 0..1

8.2.3.3 Object Properties

Property: refersTo   **Range: Item**

Inverse: Item identifiers

Definition: the unique Item to which the ItemId refers.

Cardinality: 1..1

Property: registeredIn   **Range: Registry**

Definition: the Registry which assigns a specific part or product to the partId and revisionId values..

Cardinality: 1..1

8.2.3.4 Range Roles

none.

8.2.4 Class: StandardPack

Definition: The specification for a standard unit of packaging of an Item, commonly a unit of sale and a unit of shipment.

8.2.4.1 Generalizations

none.

8.2.4.2 Datatype Properties

Property: quantityPerPackage Range: Quantity

Definition: the quantity of the Item that a single StandardPack unit contains.

Cardinality: 0..1

8.2.4.3 Object Properties

Property: forItem Range: Item

Inverse: Item packaging

Definition: the (catalog) Item that is packaged in the StandardPack.

Cardinality: 1..1

Property: notes Range: Note

Inverse: Note toPackaging

Definition: Notes, particularly for handling and shipment, that apply to the StandardPack (and to Packages that conform to that specification). Such Notes are often managed formally as parts of the documentation for the Packaging, or for a category of Items or Packaging that applies to the StandardPack in question, e.g., hazardous materials.

Note: When the Note relates to a category of Item, or a category of Packaging, the same Note (individual) can apply to more than one StandardPack.

Cardinality: 0..*

8.2.4.4 Range Roles

From: Package as specifiedBy

From: ScheduleLine as packaging

8.3 Instances

none.

9 Shipments and Shipment Schedules

This section introduces all of the major concepts for managing Shipments, although it omits most of the logistics details. Figure 9 below gives an overview of the principal concepts in this section. Detailed diagrams for Shipment Schedules, Schedule Lines, Shipments and Shipment Units appear in the subsections for those classes.

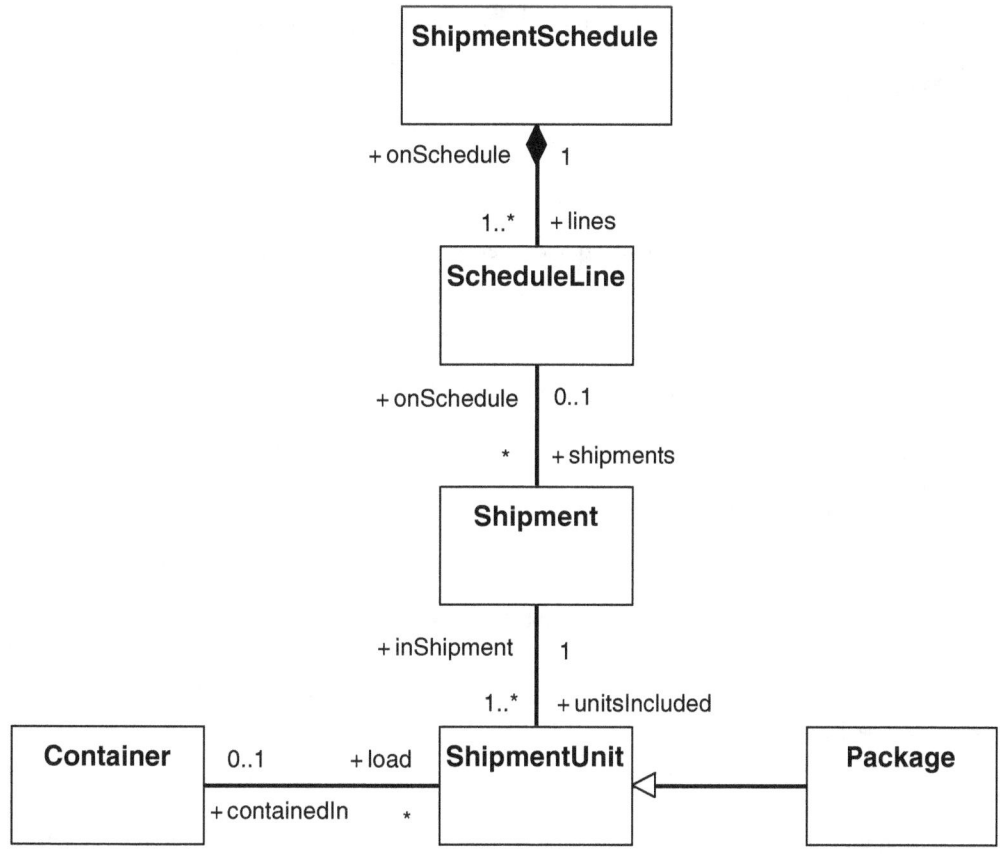

Figure 9　　Overview of Shipments and Shipment Schedules

9.1 Datatypes

9.1.1 Datatype:　ContainerTypeCode

Type: `atomic`

Definition: A Code identifying the category of Container, from UN/ECE Trade Recommendation 21

9.1.1.1 Supertypes

CodeType

9.1.1.2 Elements

none.

9.1.1.3 Range Roles

From Container as type

9.1.2 Datatype: DataIdentifierCode

Type: `atomic`

Definition: The code from ANSI MH-10.8.2 [13] that specifies what information unit the labelContent field of a PackagingLabel component identifies, such as SupplierId, Item Id, Package serial number, etc.

9.1.2.1 Supertypes

CodeType

9.1.2.2 Values

Many more values than these are specified in the code reference. These values are specifically required for certain uses in e-Kanban shipments.

Value: 1J

Definition: Unique license plate number lowest package level (unbreakable unit)

Value: 5J

Definition: Unique license plate number mixed load

Value: 6J

Definition: Unique license plate number assigned to a master load

Value: 3S

Definition: Supplier's package serial number

Value: G

Definition: Mixed handling unit

Value: M

Definition: Homogeneous handling unit

Value: S

Definition: Simplified handling unit / Inner package.

9.1.2.3 Range Roles

From MarksAndNumbers as contentType

9.1.3 Datatype: LabelPositionCode

Type: `atomic`

Definition: Code from the AIAG B-16 [14] label positioning codes for where this building block appears in the overall PackagingLabel. Values: 11Z through 17Z.

9.1.3.1 Supertypes

CodeType

9.1.3.2 Elements

none.

9.1.3.3 Range Roles

From MarksAndNumbers as positionCode

9.1.4 Datatype: MarksAndNumbersTypeCode

Type: `atomic`

Definition: The code from ANS MH10.8.3 [15] that specifies the syntax or reference term set for the labelContent of a PackagingLabel.

9.1.4.1 Supertypes

CodeType

9.1.4.2 Elements

none.

9.1.4.3 Range Roles

From MarksAndNumbers as codeType

9.1.5 Datatype: MarksAndNumbers

Type: `structure`

Definition: The values for one field (building block) of a PackagingLabel, i.e. the a label on a ShipmentUnit or its shipping container (such as a Kanban). It is derived from the Global Transport Label Standard for the Automotive Industry [14], which specifies the structure of Barcode labels and component values for Packages.

9.1.5.1 Supertypes

none.

9.1.5.2 Elements

Element: codeType   **Type: MarksAndNumbersTypeCode**

Definition: The code that specifies the syntax or reference term set for the labelContent.

Cardinality: 1..1

Element: contentType   **Type: DataIdentifierCode**

Definition: The code that specifies what information unit the labelContent field identifies.

Cardinality: 1..1

Element: labelContent   **Type:** Name

Definition: the value of the label element. This is given when the MarksAndNumbers datum represents a component of an actual label. It is usually absent when the MarksAndNumbers datum represents only the syntax of a RFID or visible label, although it may be specified if the value is constant over all actual labels of interest.

Cardinality: 0..1

Element: positionCode   **Type:** LabelPositionCode

Definition: code for where this syntactic element appears in the overall packaging label.

Cardinality: 1..1

9.1.5.3 Range   Roles

From: PackagingLabel **as elements**

9.1.6 Datatype: ScheduleType

Type: enumeration

Definition: This is a code indicating whether the Shipments under the Shipment Schedule are "Pickup based" or "Shipment based" or "Delivery based". This code reflects the business agreements for the management of the Carrier.

9.1.6.1 Supertypes

none.

9.1.6.2 Values

Value: ShipmentBased

Definition: The shipment-based ScheduleType means that for each Shipment, the carrier must leave the shipping dock within the shippingPeriod for the Shipment. For this schedule type, either the Customer or Supplier may hold the formal contract with the Carrier, but the Customer dictates the window for Shipments from the Supplier site, and the Supplier has some degree of management of the Carrier's schedule for actual pickups on the Supplier site.

Value: PickupBased

Definition: The pickup-based ScheduleType means that the customer manages the carrier and will send the carrier to the supplier site to pickup the Kanban units. Accordingly, the supplier should have Kanbans ready for pick up at the pick-up point within the pickupPeriod specified for each Shipment.

Value: DeliveryBased

Definition: The delivery-based ScheduleType means that the Supplier controls the Carrier and determines the necessary lead time for Shipment. For each Shipment, the customer dictates the deliveryPeriod during which the shipment must reach the specified delivery location. Working with the Carrier, the Supplier determines the corresponding shippingPeriod.

9.1.6.3 Range   Roles

From ShipmentSchedule **as type**

9.2 Classes

9.2.1 Class: Container

Definition: A physical container used to hold or convey ShipmentUnits or Packages of product. In general, containers are only significant in two cases:

• when the Container is reusable, has an externally readable label that identifies it, and is tracked when empty as well as when it is full. (This case applies to trailers, rail cars, freight containers, etc.)

• when the Container contains multiple ShipmentUnits corresponding to one or more separate Shipments, e.g., Shipments of different Items. (In this case, the Container is the physical unit delivered, and, even when it is disposable, it may have summary labelling and/or attached documents.)

Note: The properties of Container are depicted in Figure 13.

9.2.1.1 Generalizations

none.

9.2.1.2 Datatype Properties

Property: containerId **Range:** Name

Definition: An identifying label on the physical container.

Cardinality: 0..1

Property: RFID **Range:** Name

Definition: Radio Frequency Identifier. It is the identifying code for a container or a shipment unit of product that is the response to an RF "ping" (interrogation) of its physical RFtag device./

Cardinality: 0..1

Property: sealId **Range:** Name

Definition: The identification (label) on a Container seal. SealID is used for security purposes to ensure that the shipment is original as shipped. A seal that is different, or not intact, may trigger an alert.

Cardinality: 0..1

Property: type **Range:** ContainerTypeCode

Definition: The standard category to which the Container belongs.

Cardinality: 0..1

9.2.1.3 Object Properties

Property: load **Range:** ShipmentUnit

Inverse: ShipmentUnit.containedIn

Definition: the current contents of this Container, if any. A Container may contain ShipmentUnits that are part of different Shipments.

Cardinality: 0..*

Property: containedIn **Range:** Container

Inverse: Container.contains

Definition: the larger Container, if any, in which this Container is shipped.

Cardinality: 0..1

Property: contains **Range: Container**

Inverse: Container containedIn

Definition: the smaller Containers, if any, that are physically contained in this Container.

Cardinality: 0..*

9.2.1.4 Range Roles

none.

9.2.2 Class: Package

Definition: A Package is a physical unit of product, as packaged for storage or shipment. It a ShipmentUnit that conforms to a StandardPack(age) specification

Note: The properties of Package are depicted in Figure 13.

9.2.2.1 Generalizations

ShipmentUnit

9.2.2.2 Datatype Properties

Property: actualQuantity **Range: Quantity**

Definition: the quantity of product (Item) that the Package actually contains. When this information unit is not specified, the quantity is that given by the StandardPack specification.

Cardinality: 0..1

9.2.2.3 Object Properties

Property: specifiedBy **Range: StandardPack**

Definition: the packaging specification for the Package (the physical unit of product/material).

Cardinality: 1..1

Property: notes **Range: Note**

Definition: any Notes that explicitly apply to the physical package instance.

Cardinality: 0..*

9.2.2.4 Range Roles

none.

9.2.3 Class: PackagingLabel

Definition: The specification for the structure of data units that serve to uniquely identify a Package (Kanban unit). The structure has a well-defined physical representation, corresponding to a particular labeling technology (e.g., Barcode, RFID) and the representation standard for that technology. The model of a PackagingLabel is a set of MarksAndNumbers specifications, each of which describes one field of the PackagingLabel. A PackagingLabel specification always specifies the sequence and nature of the label fields. It may specify the content values for some or all of the fields. When all the content values are specified, the PackagingLabel is a model of one specific label that is the identifier for (and *on*) a single Package.

Note: The properties of PackagingLabel are depicted in.Figure 10 and in Figure 13.

9.2.3.1 Generalizations

none.

9.2.3.2 Datatype Properties

Property: elements   **Type: MarksAndNumbers**

Definition: the relationship between a PackagingLabel and the elements that represent the required fields.

Cardinality: 1..*

9.2.3.3 Object Properties

none.

9.2.3.4 Range Roles

From ShipmentUnit as label

From ScheduleLine as labelingRule

9.2.4 Class: ReceiptDiscrepancy

Definition: Provides receiving discrepancy information for a Shipment (such as a variance from the schedule, or from the ShipmentNotification, incorrect items, incorrect quantities, unacceptable items, etc.), although it may cite a problem with an individual ShipmentUnit (e.g., a Kanban package or a group of packages). In rare cases, a ReceiptDiscrepancy applies to a received ShipmentUnit (per the label on the Container) that does not correspond to any expected Shipment from the Supplier.

Note: The properties of ReceiptDiscrepancy are depicted in Figure 11.

9.2.4.1 Generalizations

none.

9.2.4.2 Datatype Properties

Property: actualQuantity   **Range: Quantity**

Definition: the quantity of product Item actually present in the Shipment(Unit), but only if it differs from the quantity expected.

Cardinality: 0..1

Property: issues   **Range: Text**

Definition: Each issue is a textual description of one discrepancy between the ShipmentUnits actually received by the ShipToParty and the shipment notice or the shipment schedule (if no shipment notice has been received).

Cardinality: 1..*

9.2.4.3 Object Properties

Property: inShipment **Range: Shipment**

Inverse: Shipment discrepancies

Definition: The Shipment to which the Discrepancy report is related.

Cardinality: 1..1

Property: inUnit **Range: ShipmentUnit**

Definition: The specific ShipmentUnit to which the Discrepancy report is related, if any.

Cardinality: 0..1

9.2.4.4 Range Roles

none.

9.2.5 Class: ScheduleLine

Definition: a Line Item in a ShipmentSchedule. It describes one or more Shipments of the same Item.

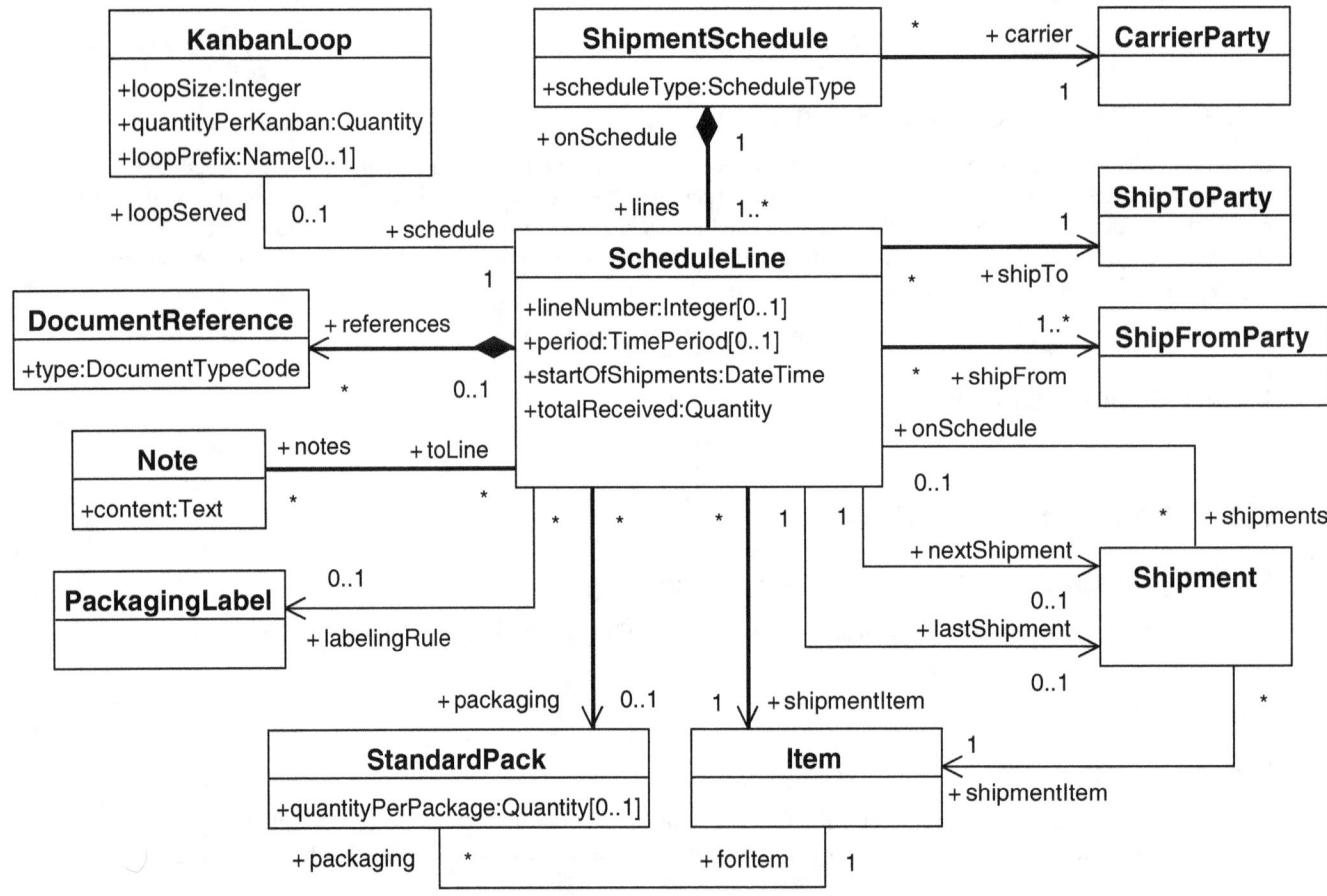

Figure 10   Schedule Lines

9.2.5.1 Generalizations

none.

9.2.5.2 Datatype Properties

Property: lineNumber   **Range:** Integer

Definition: A number that uniquely identifies the ScheduleLine within the ShipmentSchedule. The number is not required when the Schedule has only one line, or when the convention is to identify the Line by the Item shipped.

Cardinality: 0..1

Property: period   **Range:** TimePeriod

Definition: The specific time interval (first date, last date) in which Shipments corresponding to the ScheduleLine will occur.

Cardinality: 0..1

Property: startOfShipments   **Range:** DateTime

Definition: The date on which the first Shipment corresponding to the ScheduleLine occurs. This can be a future (scheduled) date or a past date, depending on whether the first Shipment has occurred or not.

Cardinality: 1..1

Property: totalReceived   **Range:** Quantity

Definition: the total quantity of product that has been shipped to date, also called "cumulative quantity".

Cardinality: 1..1

9.2.5.3 Object Properties

Property: labelingRule   **Range:** PackagingLabel

Definition: A PackagingLabel that represents the instructions for labeling the ShipmentUnits that are shipped under this ScheduleLine. In general, the PackagingLabel identifies all of the required fields and their ordering/placement, but only some of their values. The values of the remaining fields are determined by the ShipFrom facility at the time of shipment.

Cardinality: 0..1

Property: lastShipment   **Range:** Shipment

Definition: the last Shipment actually made (shipped) according to this ScheduleLine. A Shipment becomes the lastShipment when it is shipped, and remains the lastShipment until its successor is shipped.

Cardinality: 0..1

Property: loopServed   **Range:** KanbanLoop

Inverse: KanbanLoop.schedule

Definition: the KanbanLoop, if any, that is implemented by this ScheduleLine.

Cardinality: 0..1

Property: nextShipment   **Range:** Shipment

Definition: the next Shipment to be shipped according to this ScheduleLine. This relationship comes into existence when the Shipment object has been created to capture its scheduling details and it becomes the next shipment, by virtue of its predecessor becoming the lastShipment. These two events can occur in either order.

Cardinality: 0..1

Property: notes   **Range:** Note

Inverse: Note.toLine

Definition: Any textual Notes that apply to this ScheduleLine.

Cardinality: 0..*

Property: onSchedule   **Range:** ShipmentSchedule

Inverse: ShipmentSchedule.lines

Definition: the ShipmentSchedule to which the ScheduleLine belongs.

Cardinality: 1..1

Property: packaging   **Range:** StandardPack

Definition: The packaging specification, if any, that applies to Shipments against this ScheduleLine.

Cardinality: 0..1

Property: references   **Range:** DocumentReference

Definition: the Documents associated with the ScheduleLine, represented by DocumentReferences in/on/with the ScheduleLine.

Cardinality: 0..*

Property: shipFrom   **Range:** ShipFromParty

Definition: the Supplier Facility from which all materials shipped under this ScheduleLine will be shipped. Technically, there can be more than one such Facility, but in most cases, there will be only one.

Cardinality: 1..*

Property: shipments   **Range:** Shipment

Inverse: Shipment.onSchedule

Definition: the set of Shipments that correspond to this ScheduleLine. At any given time, this set consists of the historical shipments, plus any future logical Shipments for which shipping details have been decided and captured.

Cardinality: 0..*

Property: shipTo   **Range:** ShipToParty

Definition: the Party Facility that is to receive all the materials shipped under this ScheduleLine. A ScheduleLine always applies to a unique ShipTo Party.

Cardinality: 1..1

9.2.5.4 Range Roles

none.

9.2.6 Class: ShipFromParty

Definition: A PartyRole, played by a Party representing a Supplier facility, that functions as the Shipper in a KanbanLoop.

9.2.6.1 Generalizations

E-KanbanPartyRole

9.2.6.2 Datatype   Properties

none.

9.2.6.3 Object   Properties

none.

9.2.6.4 Range   Roles

From: ShipmentSchedule **as shipFrom**

From: Shipment **as shipFrom**

From: ScheduleLine **as shipFrom**

9.2.7 Class: Shipment

Definition: a set of ShipmentUnits of one Item that are conveyed simultaneously to a single receiver destination.

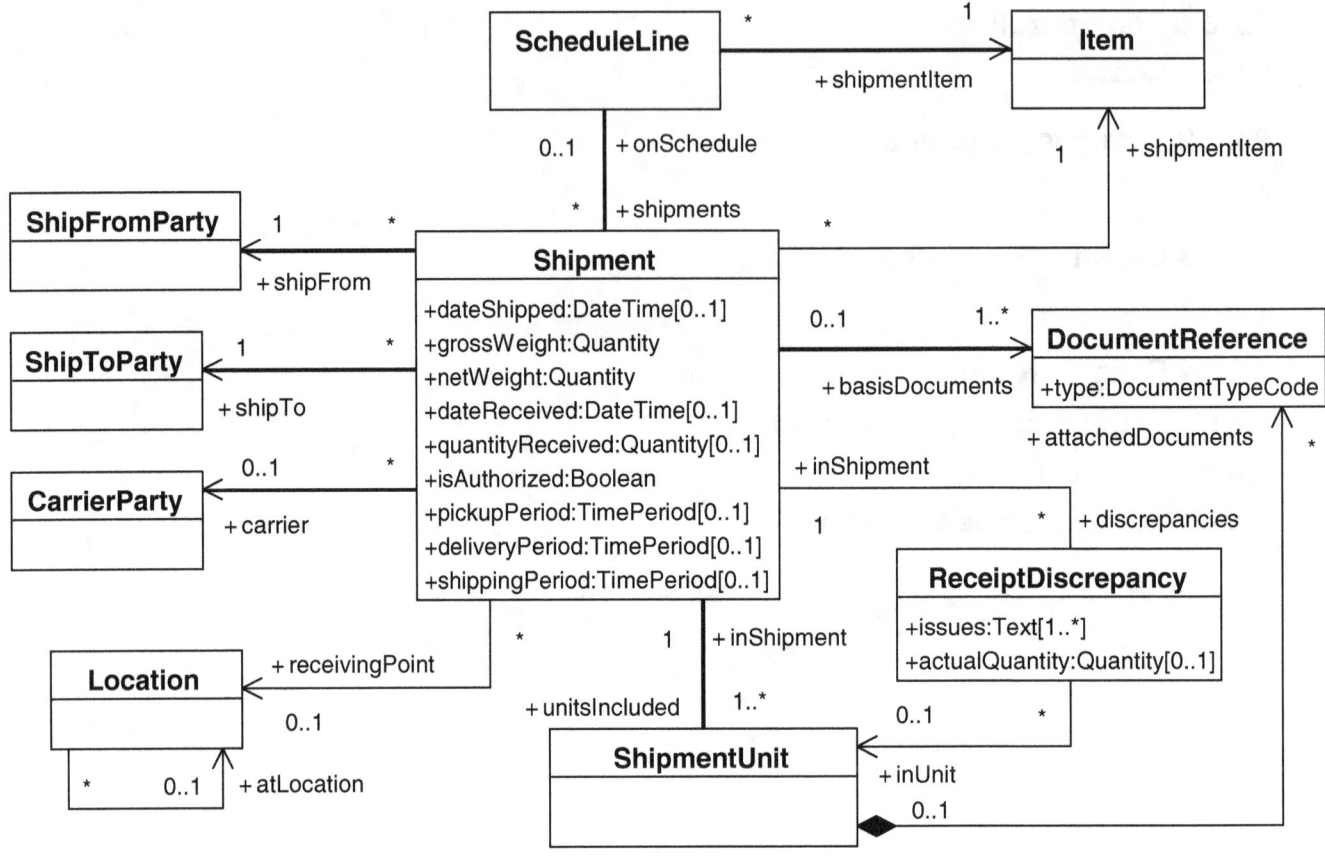

Figure 11   Shipments

9.2.7.1 Generalizations

none.

9.2.7.2 Datatype   Properties

Property: dateReceived   Range: <u>DateTime</u>

Definition: the date (and time) the customer actually receives the Shipment. The e-Kanban business process specification leaves open the exact definition. It could be the time at which the Shipment is unloaded to the receiving dock, or when the ShipmentUnits (Kanban units) are unloaded at the dock, or when the Items are unpacked, inspected and accepted, or possibly as late as when the packages are broken and the Items are fed to the line. This field is often used for performance calculations.

Cardinality: 0..1

Property: dateShipped   Range: <u>DateTime</u>

Definition: the date (and time) at which the Shipment left the Supplier facility by carrier.

Cardinality: 0..1

Property: deliveryPeriod   **Range: TimePeriod**

Definition: The time window within which the Shipment carrier must arrive at the shipTo site. This specification applies only to DeliveryBased Schedules.

Cardinality: 0..1

Property: grossWeight   **Range: Quantity**

Definition: The weight of the set of ShipmentUnits, including containers and packaging.

Cardinality: 1..1

Property: isAuthorized   **Range: Boolean**

Definition: True if this Shipment has been authorized, False if it is pending authorization.

Cardinality: 1..1

Property: netWeight   **Range: Quantity**

Definition: the weight of product (Items) contained in the Shipment.

Cardinality: 1..1

Property: pickupPeriod   **Range: TimePeriod**

Definition: The time window within which the Shipment must be available for pickup at the shipFrom site. This specification applies only to PickupBased Schedules.

Cardinality: 0..1

Property: quantityReceived   **Range: Quantity**

Definition: the quantity of product actually received at the Customer facility (when the Shipment was unpacked). The way in which this value is reckoned may be affected by acceptance rules for things like damaged containers, but properly this value should be the quantity of the Item that arrived, as distinct from the quantity that was accepted.

Cardinality: 0..1

Property: shippingPeriod   **Range: TimePeriod**

Definition: The time window within which the Shipment carrier must leave the shipFrom site. Depending on the ScheduleType, this may represent a Customer-provided specification, or a Supplier-derived specification, based on the specified deliveryPeriod. This specification applies only to ShipmentBased and DeliveryBased Schedules.

Cardinality: 0..1

9.2.7.3 Object Properties

Property: discrepancies   **Range: ReceiptDiscrepancy**

Inverse: ReceiptDiscrepancy.inShipment

Definition: Any discrepancies in this Shipment, as reported by the receiver (ShipToParty).

Cardinality: 0..*

Property: receivingPoint   **Range: Location**

Definition: the particular unloading zone for the Shipment, if specified. This is usually provided as an addendum to the facility Location associated with the ShipToParty.

Cardinality: 0..1

Property: transportedIn Range: Equipment

Definition: the Equipment item that is used to convey the Shipment. This relationship is only captured when the Equipment, such as a trailer or rail car, is left at the Supplier or Customer site. In such cases, the Supplier or Customer assumes some responsibility for the Equipment and must have some arrangement with the EquipmentOwnerParty. In a few cases, the Equipment is identified in order to facilitate access to the customer site.

Cardinality: 0..1

Property: shipTo Range: ShipToParty

Definition: the Customer facility to which the Shipment is to be delivered.

Cardinality: 1..1

Property: shipFrom Range: ShipFromParty

Definition: the Supplier facility at which the Shipment is to be picked up, or from which it is to be shipped.

Cardinality: 1..1

Property: carrier Range: CarrierParty

Definition: the Carrier that transports the shipment, or provides primary logistics management for the Shipment. By convention, the Logistics Service Bureau/Agent is used for the Carrier when there are multiple actual transporters and other agents involved. The Carrier for a Shipment always exists, but the Carrier may not be identified in the transactions between the trading partners.

Cardinality: 0..1

Property: onSchedule Range: ScheduleLine

Inverse: ScheduleLine.Shipments

Definition: the relationship between a Shipment and the ScheduleLine to which it corresponds. Every Shipment should correspond to exactly one ScheduleLine. This relationship is said to be optional in order to deal with (reports of) erroneous Shipments, whose association to the proper ScheduleLine may require a recovery process.

Cardinality: 0..1

Property: basisDocuments Range: DocumentReference

Definition: Document references to

- contractual documents that define or relate to the business agreement under which the Shipment takes place, such as a Purchase Order, a Kanban contract, a shipment authorization, or a Bill of Lading.

- contractual documents and licenses that apply to the conveyance of the goods, such as carrier and logistics agreements, import/export licenses, inspection documents, etc.

- physical documents that go with the Shipment, but apply to the conveyance, or to the entire set of shipments conveyed at once, such as a Manifest or a Bill of Lading for a consignment of Shipments of several Items.

Cardinality: 1..*

Property: unitsIncluded Range: ShipmentUnit

Inverse: ShipmentUnit.inShipment

Definition: the ShipmentUnits that are part of the Shipment.

Cardinality: 1..*

An Ontology for the e-Kanban Business Process NISTIR 7404

Property: shipmentItem   Range: Item

Definition: the product or material Item that is being supplied in this Shipment. A logical Shipment contains units of exactly one Item.

Cardinality: 1..1

9.2.7.4 Range Roles

From: ScheduleLine as lastShipment

From: ScheduleLine as nextShipment

From: ReceiveDeliveryNotification as reportsOn

From: ShipmentNotification as reportsOn

9.2.8 Class: ShipmentSchedule

Definition: A BusinessObjectDocument that describes all of the Shipments for some collection of Items that are supplied by one Supplier to one Customer under a common set of replenishment rules. The relationship is established by one or more business agreements between the two Parties that identify the Items to be provided, the ShipFrom and ShipTo Facility locations, the business rules for the replenishment interactions, and the assignment of logistics responsibilities. The ShipmentSchedule contains one ScheduleLine per (combination of) Item, ShipFrom location and ShipTo location covered by the agreement(s). All Items are supplied under a common set of replenishment rules, with a common assignment of responsibility for the movement of the goods.

In the e-Kanban business process, the common set of replenishment rules is a specialization of the IV&I e-Kanban business process.

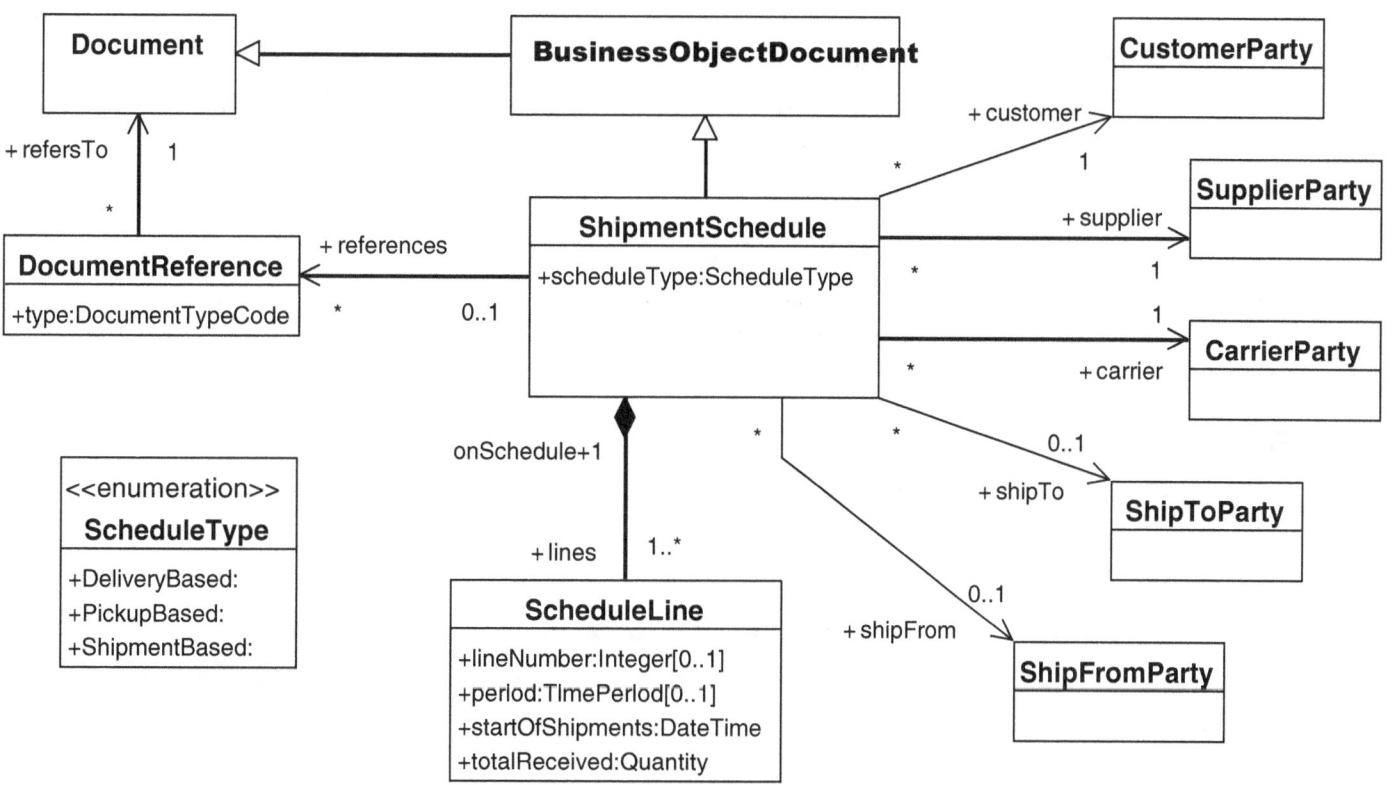

Figure 12   ShipmentSchedules

57

9.2.8.1 Generalizations

BusinessObjectDocument

9.2.8.2 Datatype Properties

Property: scheduleType Range: ScheduleType

Definition: The type of shipping arrangement covered by the Schedule. See ScheduleType for details.

Cardinality: 1..1

9.2.8.3 Object Properties

Property: carrier Range: CarrierParty

Definition: The primary logistics service provider (LSP) for all Shipments corresponding to the ShipmentSchedule. As discussed in 10.2.1, the term Carrier is used in the e-Kanban business process to refer to the Party contractually responsible for the movement of goods under the business agreement(s) to which the ShipmentSchedule corresponds. When the same Customer and Supplier parties are involved, but the logistics responsibilities are different, the ShipmentSchedules must be different.

Cardinality: 1..1

Property: customer Range: CustomerParty

Definition: The Customer to which this ShipmentSchedule applies. A ShipmentSchedule always applies to a particular Customer-Supplier relationship and therefore to exactly one CustomerParty.

Cardinality: 1..1

Property: supplier Range: SupplierParty

Definition: The Supplier to which this ShipmentSchedule applies. A ShipmentSchedule always applies to a particular Customer-Supplier relationship and therefore to exactly one SupplierParty.

Cardinality: 1..1

Property: shipTo Range: ShipToParty

Definition: the destination (Facility) for all Shipments under this ShipmentSchedule. This relationship does not exist when multiple Customer Facilities are involved in the ShipmentSchedule.

Cardinality: 0..1

Property: shipFrom Range: ShipFromParty

Definition: the supplier Facility for all Shipments under this ShipmentSchedule. This relationship does not exist when multiple Supplier Facilities are involved in the ShipmentSchedule.

Cardinality: 0..1

Property: references Range: DocumentReference

Definition: the Documents associated with the ShipmentSchedule, represented by DocumentReferences in/on/with the schedule.

Cardinality: 0..*

Property: lines Range: ScheduleLine

Inverse: ScheduleLine.onSchedule

Definition: the ScheduleLines contained in the ShipmentSchedule.

Cardinality: 1..*

9.2.8.4 Range Roles

From ~~SyncShipmentSchedule~~ as BOD

9.2.9 Class: ShipmentUnit

Definition: A physical packaged unit of one or more Items that is contained in a Shipment. A ShipmentUnit can consist of smaller ShipmentUnits. A ShipmentUnit is all or part of the "load" in a Container. Each such unit can have separately attached documents, and it may nominally attach documents from a higher-level of packaging.

Note: A ShipmentUnit can be a Package, a group of Packages, or possibly an odd number of parts stuffed into an ad hoc container. A ShipmentUnit can be a carload of palettes of boxes of parts.

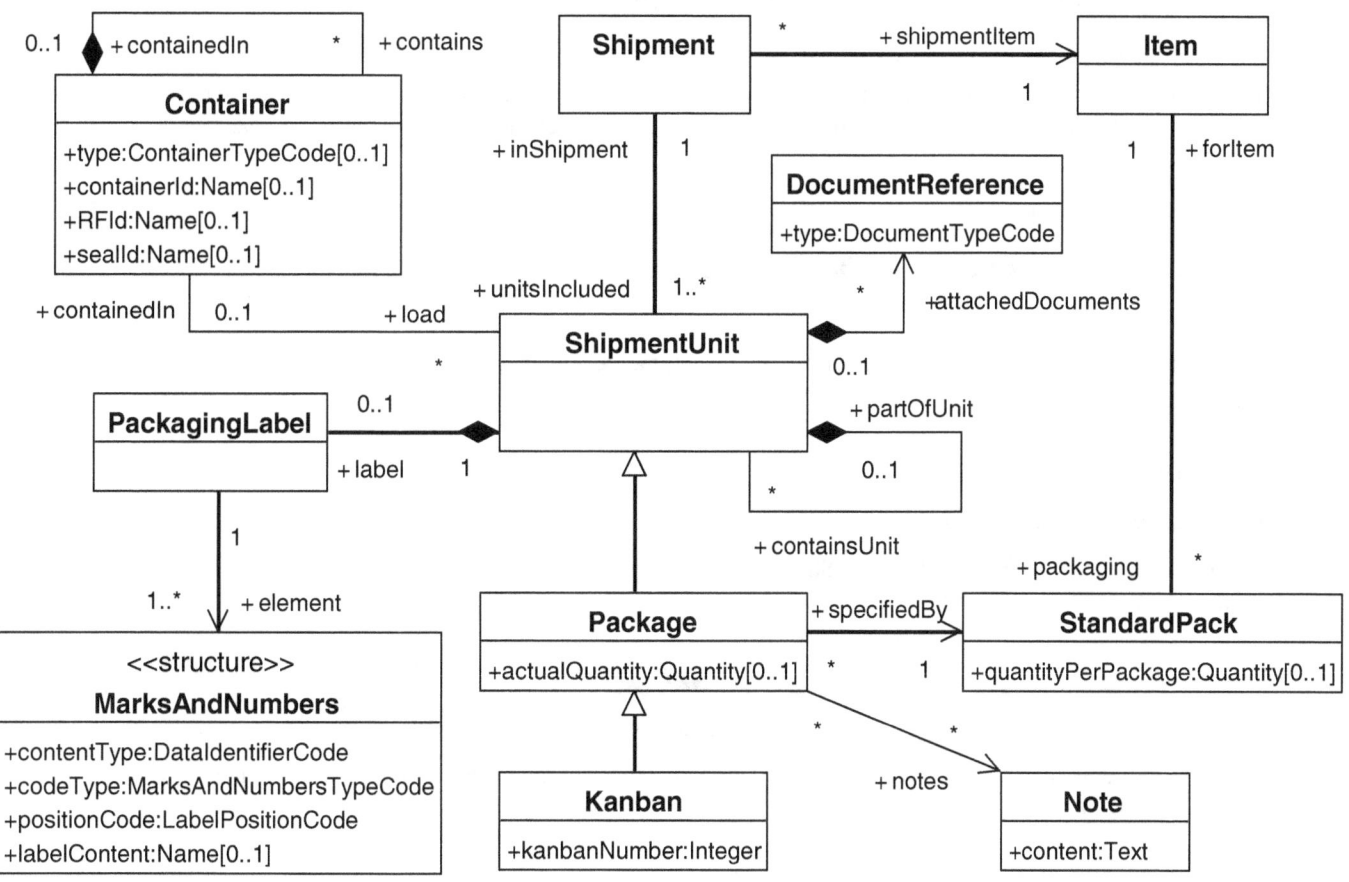

Figure 13   Shipment Units

9.2.9.1 Generalizations

none.

9.2.9.2 Datatype Properties

none

9.2.9.3 Object Properties

Property: attachedDocuments Range: DocumentReference

Definition: Document references may be to a physical document that goes with the shipment such as Manifest or Bill of Lading, or to the BlanketPO/Release. If the shipment is across a border, a Manifest is required, and Bills of Lading may be required. If the trucker picks up parts form 3 locations, he has 3 Bills of Lading. The Manifest is the summary of the BOLs. The trucker may or may not summarize the BOLs into a single Manifest. These are like loading and unloading documents. This document reference can override the document reference of the same "type" that is specified in the header section. The document reference here must not have the type that refer to the SyncShipmentSchedule document because there can be only one instance of shipment for a shipment schedule authorization in the e-Kanban business process.

Cardinality: 0..*

Property: containedIn   Range: Container

Inverse: Container.load

Definition: the Container in which this ShipmentUnit is shipped. This relationship is optional, because the Container may be an integral part of the ShipmentUnit, or may be expendable dunnage that is not managed as a Container.

Cardinality: 0..1

Property: containsUnit   Range: ShipmentUnit

Inverse: ShipmentUnit.partOfUnit

Definition: the smaller ShipmentUnits, if any, that this ShipmentUnit contains.

Cardinality: 0..*

Property: inShipment   Range: Shipment

Inverse: Shipment.unitsIncluded

Definition: the Shipment that contains the ShipmentUnit.

Cardinality: 1..1

Property: label   Range: PackagingLabel

Definition: An externally readable, possibly structured, label on a ShipmentUnit or Package.

Note: The PackagingLabel for a given ShipmentUnit specifies the complete content of the label, including the values for all fields.

Cardinality: 0..1

Property: partOfUnit   Range: ShipmentUnit

Inverse: ShipmentUnit.containsUnit

Definition: The larger ShipmentUnit, if any, that this ShipmentUnit is part of.

Cardinality: 0..1

9.2.9.4 Range Roles

From: ReceiptDiscrepancy **as:** ReceiptDiscrepancy.inUnit

9.2.10 Class: ShipToParty

Definition: A PartyRole, played by a Party representing a Customer facility, that is the receiver of the Kanbans in a KanbanLoop.

9.2.10.1 Generalizations

E-KanbanPartyRole

9.2.10.2 Datatype Properties

none.

9.2.10.3 Object Properties

none.

9.2.10.4 Range Roles

From: ShipmentSchedule **as shipTo**

From: ScheduleLine **as shipTo**

From: Shipment **as shipTo**

9.3 Instances

none.

10 Carriers and Equipment

This section contains the limited model of Carrier Parties and the management of transportation equipment that is needed for the e-Kanban process as specified by the AIAG. The concepts are depicted in Figure 14 and described below.

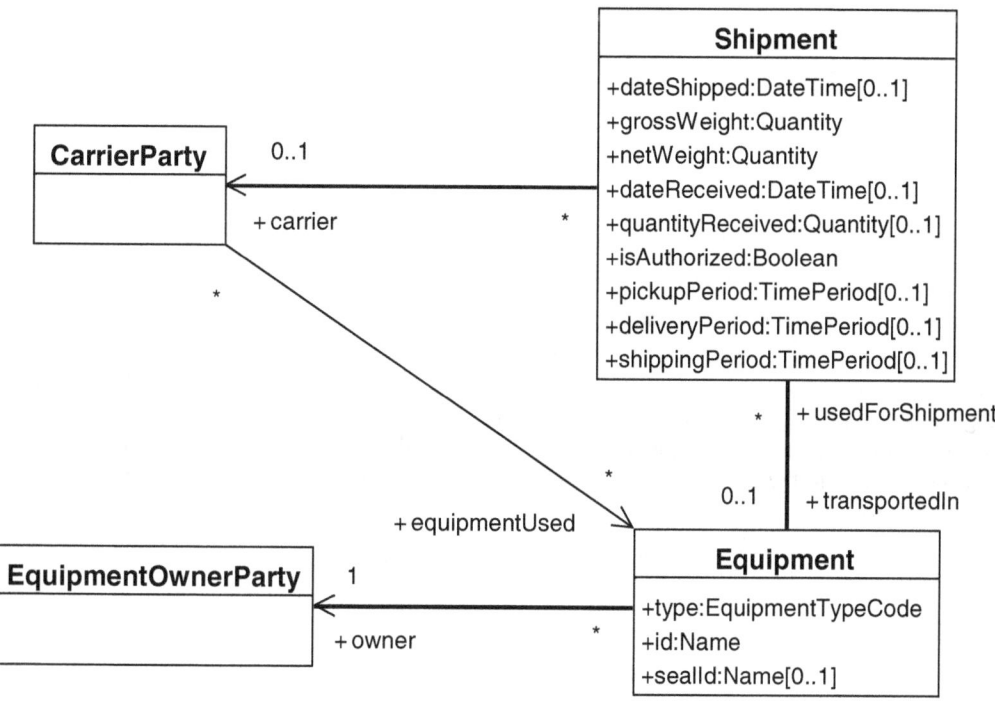

Figure 14   Carriers and Equipment

10.1 Datatypes

10.1.1 Datatype: EquipmentTypeCode

Type: atomic

Definition: A code from EDIFACT Code 8053 [16] designating the type of the equipment:

10.1.1.1 Supertypes

CodeType

10.1.1.2 Values

The following values are in use in the IV&I business process.

Value AQ

Definition: Road/rail trailer. Trailer designated for combined road/rail use.

Value BPP

Definition: Truck and trailer combination being transported: a road vehicle that is capable of carrying goods, has an attached trailer, and is being carried on another means of transport.

Value TE

Definition: Trailer A vehicle without motive power, designed for the carriage of cargo and to be towed by a motor vehicle.

10.1.1.3 Range Roles

From Equipment **as type**

10.2 Classes

10.2.1 Class: CarrierParty

Definition: The Party responsible for movement of items between the Supplier and the Customer. The Party *responsible* for the movement of items is not necessarily the transporter. In general, the movement of goods may involve more than one transporter, and it may involve other logistical services. The intent of CarrierParty is the Party serving as the prime "Logistics Service Provider", which may be the Customer, the Supplier, or a third party. The parties that perform the individual logistics services are not modeled in the e-Kanban business process – they are seen as subcontractors to the CarrierParty. In some cases, all (the only) logistics services provided are provided by the CarrierParty directly.

In the e-Kanban business process, the Carrier:
- Performs the delivery function
- May be employed or contracted by the customer or the supplier
- May monitor visibility tool for replenishment, in-transit, consumption, and alerts

10.2.1.1 Generalizations

E-KanbanPartyRole

10.2.1.2 Datatype Properties

none.

10.2.1.3 Object Properties

Property: equipmentUsed   Range: Equipment

Definition: the relationship between the Carrier and the Equipment items used. Note that this relationship is purely documentary, in that the interesting relationship is Equipment to Shipment.

Cardinality: 0..*

10.2.1.4 Range Roles

From: Shipment **as carrier**

From: ShipmentSchedule **as carrier**

10.2.2 Class: Equipment

Definition: A container, trailer, or other physical conveyance object used to hold and move goods in a material replenishment Shipment. In particular, equipment that may remain on the ShipTo (customer) site.

10.2.2.1 Generalizations

none.

10.2.2.2 Datatype Properties

Property: id   **Range: Name**

Definition: The identifier on the Equipment unit, usually in the form of a visible label.

Cardinality: 1..1

Property: sealId   **Range: Name**

Definition: The identification (label) on a Container seal. SealID is used for security purposes to ensure that the shipment is original as shipped. A seal that is different, or not intact, may trigger an alert.

Cardinality: 0..1

Property: type   **Range: EquipmentTypeCode**

Definition: A Code designating the type of the equipment, as defined in 10.1.1:
Cardinality: 1..1

10.2.2.3 Object Properties

Property: owner   **Range: EquipmentOwnerParty**

Definition: the Party that owns this item of Equipment.

Cardinality: 1..1

10.2.2.4 Range Roles

From: Shipment as transportedIn

From: CarrierParty as equipmentUsed

10.2.3 Class: EquipmentOwnerParty

Definition: A PartyRole for a Party that owns equipment (containers, trailers, etc.) that is used to hold goods in performing a Shipment, and possibly left on the Supplier site or the Customer site, in a material replenishment process. The Supplier, Customer and Carrier may all have formal relationships with the EquipmentOwnerParty, but these are not documented in the model.

10.2.3.1 Generalizations

E-KanbanPartyRole

10.2.3.2 Datatype Properties

none.

10.2.3.3 Object Properties
none.

10.2.3.4 Range Roles
From: Equipment as owner

10.3 Instances
none.

11 Kanbans

This section introduces the specific concepts that relate to the e-Kanban process, as distinct from other material replenishment business processes. There are two basic concepts: the Kanban material unit and the Kanban Loop, which is the conceptual entity being managed by the process.

The Kanban agreement between a Customer and a Supplier defines the business rules for the supply of one or more Items. There is a separate KanbanLoop for each Item supplied under the agreement. The KanbanLoop represents the agreed-on "parameters" of the e-Kanban process used for replenishment of that Item in an agreed-on set of ShipTo facilities. In operation, the Kanban agreement corresponds to a ShipmentSchedule. Each KanbanLoop corresponds to a ScheduleLine in the ShipmentSchedule. The Kanban itself is the nominal ShipmentUnit.

These concepts are depicted in Figure 15 and described below.

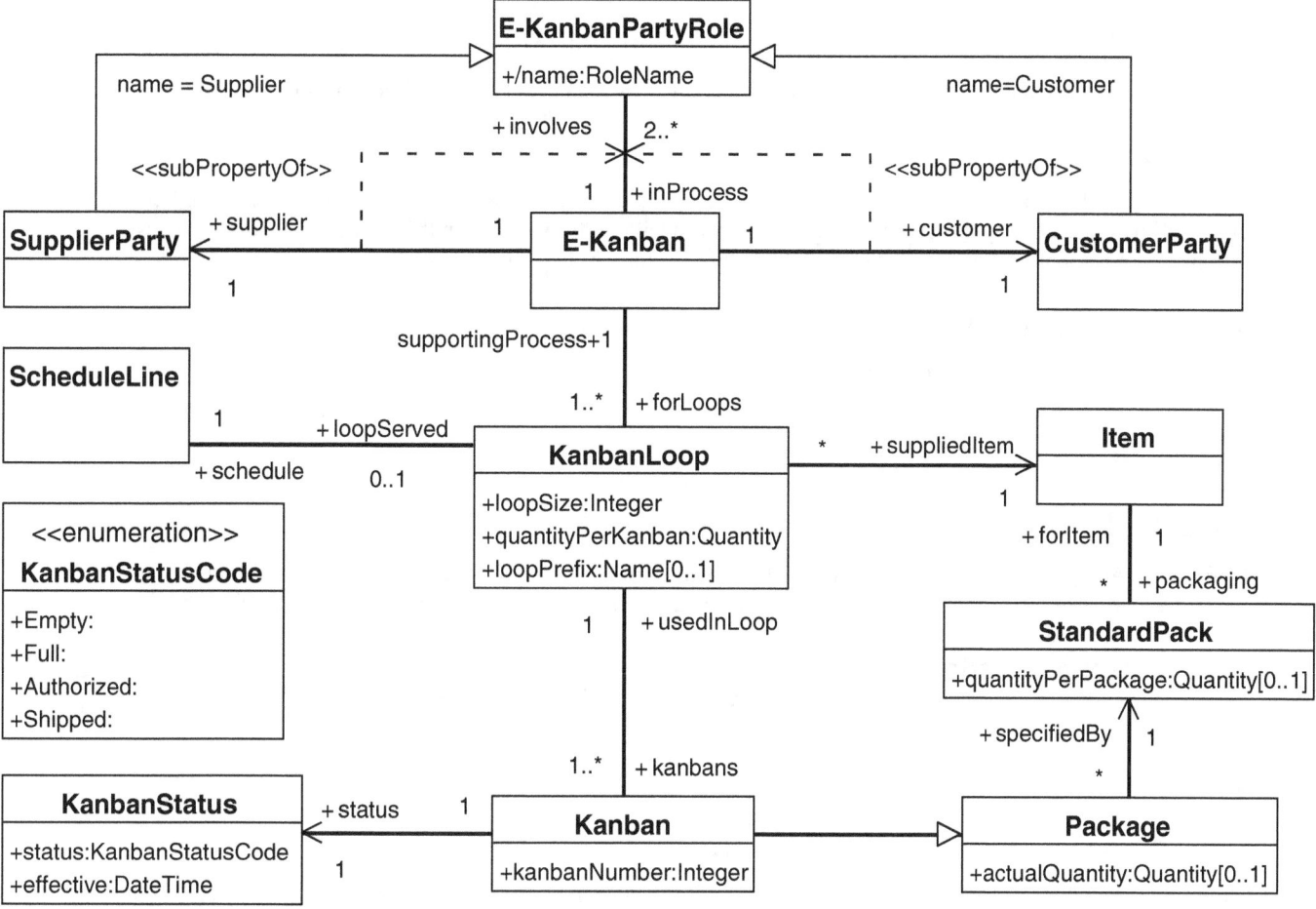

Figure 15   Kanban Concepts

11.1 Datatypes

11.1.1 Datatype: KanbanStatusCode

Type: `enumeration`

Definition: A code for the status of a Kanban unit, one of: Empty, Full, Authorized, Shipped.

11.1.1.1 Supertypes

none.

11.1.1.2 Values

Value: Authorized

Definition: An authorization (order) for replenishing this Kanban unit has been received. The Kanban unit is scheduled for shipment to the Customer.

Value: Shipped

Definition: The Kanban unit has been shipped from the Supplier to the Customer's ShipTo site, but the Customer has not acknowledged receipt and acceptance of the Kanban.

Value: Full

Definition: The Kanban unit has been received and accepted by the customer at the ShipTo facility.

Value: Empty

Definition: The Kanban unit has been consumed at the customer site. The Kanban container, if it is reusable, is now empty. If it is managed by the Supplier, the container is available for pickup. In the e-Kanban business process, designating a Kanban empty does not constitute authorization to replenish.

11.1.1.3 Range Roles

From KanbanStatus **as status**

11.2 Classes

11.2.1 Class: Kanban

Definition: A physical unit of shipment that is the reference unit of a product (item) for shipment and consumption in the Kanban business process. Every Kanban is a Package – each Kanban conforms to a StandardPack.

11.2.1.1 Generalizations

Package

11.2.1.2 Datatype Properties

Property: kanbanNumber   **Range:** Integer

Definition: a counter that identifies the Kanban in the KanbanLoop series. In some e-Kanban processes, the counter is "absolute" – counted from the first shipment. In others, it is the "relative" position in the KanbanLoop, and therefore never exceeds the loopSize value.

Cardinality: 1..1

11.2.1.3 Object Properties

Property: status   **Range:** KanbanStatus

Definition: the current (most recent) status of the Kanban.

Cardinality: 1..1

Property: usedInLoop **Range:** KanbanLoop

Inverse: KanbanLoop.kanbans

Definition: the KanbanLoop to which this Kanban belongs.

Cardinality: 1..1

11.2.1.4 Range Roles

From: KanbanConsumption **as reportsOn**

11.2.2 Class: KanbanLoop

Definition: the unit of management in an e-Kanban process. The KanbanLoop represents the agreement for supplying a particular Item. Its properties represent the agreed-on "parameters" of the e-Kanban process used for replenishment of that Item in an agreed-on set of ShipTo facilities.

Note: Kanban loops may or may not have explicit identifiers. They can be explicitly identified by a loopPrefix (see below), or implicitly identified by Item (id) and the PartyIds of some or all of the Customer, Supplier, ShipTo, and ShipFrom Parties.

11.2.2.1 Generalizations

none.

11.2.2.2 Datatype Properties

Property: loopPrefix **Range:** Name

Definition: an explicit customer-specified identifier for the Kanban loop, used as a prefix to the Kanban number in the labels on Kanban packages. In some e-Kanban processes, no such prefix is used.

Cardinality: 0..1

Property: loopSize **Range:** Integer

Definition: the number of Kanban containers that circulate in the KanbanLoop.

Cardinality: 1..1

Property: quantityPerKanban **Range:** Quantity

Definition: the amount of the supplied Item that is conveyed in a single Kanban unit (container).

Cardinality: 1..1

11.2.2.3 Object Properties

Property: kanbans **Range:** Kanban

Inverse: Kanban.usedInLoop

Definition: the Kanban packages that belong to this KanbanLoop

Cardinality: 1..*

Property: schedule **Range:** ScheduleLine

Inverse: ScheduleLine.loopServed

Definition: the ScheduleLine that corresponds to the supply schedule implementing this KanbanLoop.

Cardinality: 1..1

Property: suppliedItem   **Range:** Item

Definition: the product/material Item that is supplied via this KanbanLoop. The ItemId may be part of the identifier for the KanbanLoop itself.

Cardinality: 1..1

Property: supportingProcess   **Range:** E-Kanban

Inverse: E-Kanban.forLoops

Definition: the e-Kanban process instance that operates the KanbanLoop.

Cardinality: 1..1

11.2.2.4 Range Roles

none.

11.2.3 Class: KanbanStatus

Definition: the most recent status of the Kanban (shipment unit). The information unit includes both the actual status code and the date/time at which it became the status of the Kanban.

11.2.3.1 Generalizations

none.

11.2.3.2 Datatype Properties

Property: effective   **Range:** DateTime

Definition: represents the date and time as of which the status/value changed.

Cardinality: 1..1

Property: status   **Range:** KanbanStatusCode

Definition: The current status of the Kanban unit, one of: Empty, Full, Authorized, Shipped.

Cardinality: 1..1

11.2.3.3 Object Properties

none.

11.2.3.4 Range Roles

From: Kanban **as status**

11.3 Instances

none.

12 Messages

This section introduces the Message concept, as it is defined by the Open Applications Group. It also defines the four types of message that are specified in the IV&I e-Kanban interchange specification. The OAG Message concept is depicted in Figure 16 and described in detail below. The four e-Kanban message types are depicted in the subsections that contain their class definitions.

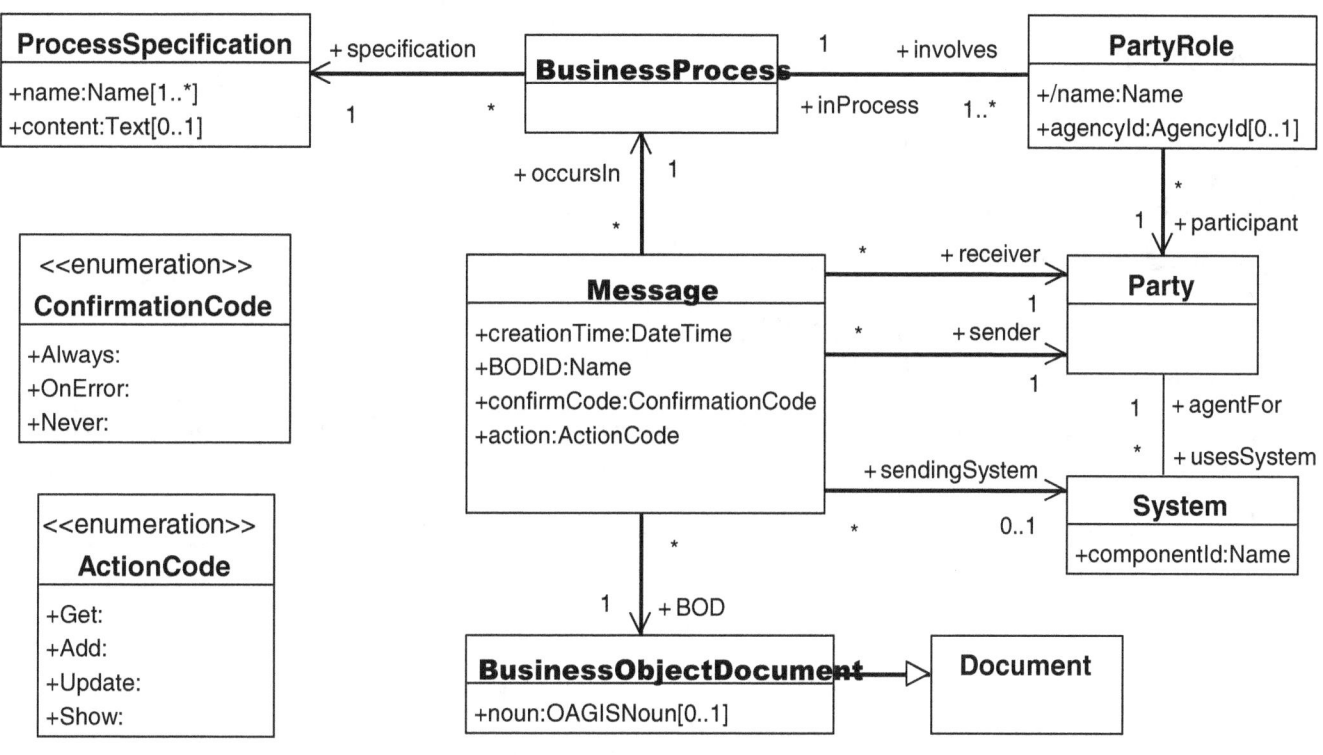

Figure 16   Messages

12.1 Datatypes

12.1.1 Datatype: ActionCode

Type: enumeration

Definition: The OAG standard "verb" for the action of this message on the referenced BusinessObjectDocument: one of Add, Get, Show, Update.

12.1.1.1 Supertypes

none.

12.1.1.2 Values

Value: Add

Definition: Indicates that the Message provides a new or revised BusinessObjectDocument for which the information is maintained by the Receiver Party, or in the case of "Sync" messages, by both parties.

Value: Get

Definition: Indicates that the Message is a request from the Sender for information maintained by the Receiver. The "BusinessObjectDocument" contained in the Message is just skeleton of identifiers and subject matter constraints that the Receiver can use to identify the BusinessObjectDocument that is requested.

Value: Show

Definition: Indicates that the Message provides a BusinessObjectDocument for which the information is maintained by the Sender Party. This is usually a response to a request by the Receiver Party.

Value: Update

Definition: Indicates that the Message provides only the new, corrected, or modified information for a BusinessObjectDocument maintained by the Receiver Party.

12.1.1.3 Range Roles

From Message as action

12.1.2 Datatype: ConfirmationCode

Type: `enumeration`

Definition: An OAGIS-defined code that indicates whether the receiver should respond to the sender on receipt of a Message.

12.1.2.1 Supertypes

none.

12.1.2.2 Values

Value: Always

Definition: Receiver should always send Confirmation Message when this message is received, parsed, and accepted. The Confirmation should come as soon as possible, and does not wait for the Message to be acted on.

Value: Never

Definition: Receiver should never send a Confirmation Message on receipt. If the Message contains an error, or the requested Action cannot be performed, the Receiver should simply discard the Message.

Value: OnError

Definition: Receiver should send a Confirmation Message when the message is received, but only if there is a problem with the Message. The Sender will assume acceptance, unless it receives an Error confirmation.

12.1.2.3 Range Roles

From Message as confirmCode

12.1.3 Datatype: OAGISNoun

Type: `atomic`

Definition: A Noun defined in the OAGIS vocabulary to be the subject of some message. It is usually the name of a type of BusinessObjectDocument.

12.1.3.1 Supertypes

Name

12.1.3.2 Elements

none.

12.1.3.3 Range Roles

From BusinessObjectDocument as noun

12.2 Classes

12.2.1 Class: BusinessObjectDocument

Definition: A Document containing a collection of information formally organized for some business purposes.

Properties: abstract

12.2.1.1 Generalizations

Document

12.2.1.2 Datatype Properties

Property: noun **Range: OAGISNoun**

Definition: A name for the type of the BusinessObjectDocument (BOD) that is used in OAGIS Messages..

Cardinality: 0..1

12.2.1.3 Object Properties

none.

12.2.1.4 Range Roles

From: Message as BOD

12.2.2 Class: Message

Definition: (OAG Message) A structured information unit that is communicated between parties to a business process.

Properties: abstract

12.2.2.1 Generalizations

none.

12.2.2.2 Datatype Properties

Property: BODID **Range: Name**

Definition: A unique identifier for the Message instance (an individual transmission of a message), set by the sender.

Cardinality: 1..1

Property: action Range: ActionCode

Definition: The OAG ActionCode for the action of this message on the BusinessObjectDocument it contains (or refers to).

Cardinality: 1..1

Property: confirmCode Range: ConfirmationCode

Definition: The OAG ConfirmationCode specifying the sender's requirements for the receiver's response on arrival of the message.

Cardinality: 1..1

Property: creationTime Range: DateTime

Definition: The date and time at which the message was created and sent by the (original) sender.

Cardinality: 1..1

12.2.2.3 Object Properties

Property: BOD Range: BusinessObjectDocument

Definition: the BusinessObjectDocument that is the business content (the payload) of the Message.

Cardinality: 1..1

Property: occursIn Range: BusinessProcess

Definition: the BusinessProcess instance in which the Message is an actual communication between the business partners.

Cardinality: 1..1

Property: receiver Range: Party

Definition: the Party that receives this Message.

Cardinality: 1..1

Property: sender Range: Party

Definition: the Party that sent this Message.

Cardinality: 1..1

Property: sendingSystem Range: System

Definition: the software System that sent this Message.

Cardinality: 0..1

12.2.2.4 Range Roles

none.

12.2.3 Class: System

Definition: An application software system that serves as the electronic agent for a Party in the e-Kanban business process.

The Customer System performs the following functions:

- Generates the material consumption (SyncKanbanConsumption) messages
- Generates the replenishment authorization (SyncShipmentSchedule) messages
- Generates the receipt and acceptance (SyncReceiveDelivery) messages

The Supplier System performs the following functions:

- Generates the shipment notification (SyncShipment) messages

The Customer Repository System, or Customer Enterprise Resource Planning (ERP) System, is the System that tracks materials inventory, materials assignments and consumption schedules, and materials orders, including the status of planned, authorized and received Kanbans.

The Supplier Repository System, or Supplier ERP System, is the System that tracks product inventory, manufacturing schedules, and product orders, including the status of planned, authorized, shipped and received Kanbans.

The Inventory Visibility Tool is the System that provides on-line visibility of Kanban inventory levels at the Customer site, enables alert capability and supports decision-making. Both the Customer and Supplier may have Inventory Visibility tools associated with, or separate from, their repository systems. The Inventory Visibility Tool performs the following functions:

- Display of Kanban status information
- Generates alerts based on status criteria

The intent of the e-Kanban business process is to support the situation in which the Customer Inventory Visibility Tool is the "Customer System" described above, and the Supplier Inventory Visibility Tool is the "Supplier System" described above. It is possible, however, that either of these is the Repository System instead of an IV Tool.

12.2.3.1 Generalizations

none.

12.2.3.2 Datatype Properties

Property: componentId   **Range: Name**

Definition: An Identifier for the application software system, in particular, the identifier for the system that actually composed and sent the Message, such as an Inventory Visibility Tool. The identifier usually takes the form of a software product name and a release version. This identifier identifies the specific software set, but not necessarily the running instance of it. The unique identification of the System also involves the .agentFor Party it represents.

Cardinality: 1..1

12.2.3.3 Object Properties

Property: agentFor   **Range: Party**

Inverse: Party usesSystem

Definition: The Party (business partner) that is represented by this System in the electronic business process. The .agentFor Party(Id) may be a part of the complete identification of the software System.

Cardinality: 1..1

12.2.3.4 Range Roles

From: Message as sendingSystem

An Ontology for the e-Kanban Business Process   NISTIR 7404

12.2.4 Class: SyncKanbanConsumption

Definition: A Message used to inform the Supplier (system) that one or more Kanban units have been consumed at the Customer site. It also informs the Supplier that the corresponding Kanban containers (if they are tracked) are empty and ready for pickup. This allows the Supplier to update its "actual demand" information, plan replenishment, and schedule container pickup if necessary.

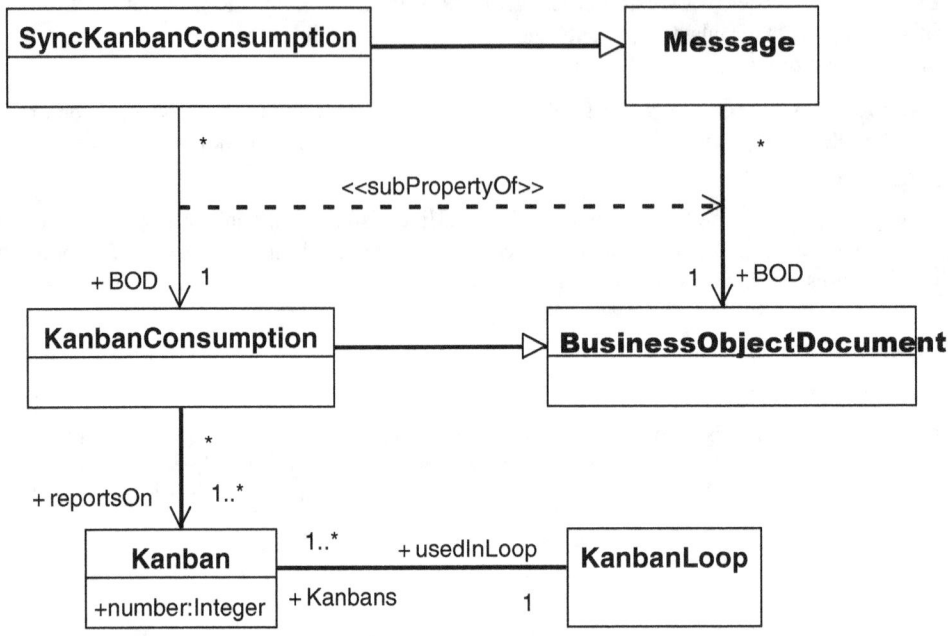

Figure 17   SyncKanbanConsumption Message

12.2.4.1 Generalizations

Message

12.2.4.2 Datatype Properties

none.

12.2.4.3 Object Properties

Property: BOD   **Range:** KanbanConsumption

SubpropertyOf: Message.BOD

Definition: The KanbanConsumption document that is the payload of the SyncKanbanConsumption message.

Cardinality: 1..1

12.2.4.4 Range Roles

none.

12.2.5 Class: KanbanConsumption

Definition: A BusinessObjectDocument that identifies the state of a one or more Kanbans (containers) as empty -- the material has been "consumed" at the Customer site.

Note: KanbanConsumption is depicted in Figure 17.

12.2.5.1 Generalizations

BusinessObjectDocument

12.2.5.2 Datatype Properties

From SyncKanbanConsumption as BOD

12.2.5.3 Object Properties

Property: reportsOn   **Range:** Kanban

Definition: the set of Kanbans that are the subject of the KanbanConsumption document.

Cardinality: 1..*

12.2.5.4 Range Roles

none.

12.2.6 Class: SyncReceiveDelivery

Definition: This Message is used by the Customer (system) to inform the Supplier (system) that one or more Shipments have been received at the Customer site. It also allows the Customer to report unexpected or unauthorized Shipments, incorrectly labeled or incorrectly routed ShipmentUnits, damaged ShipmentUnits, and other discrepancies between the expected ShipmentUnits and the actual ShipmentUnits.

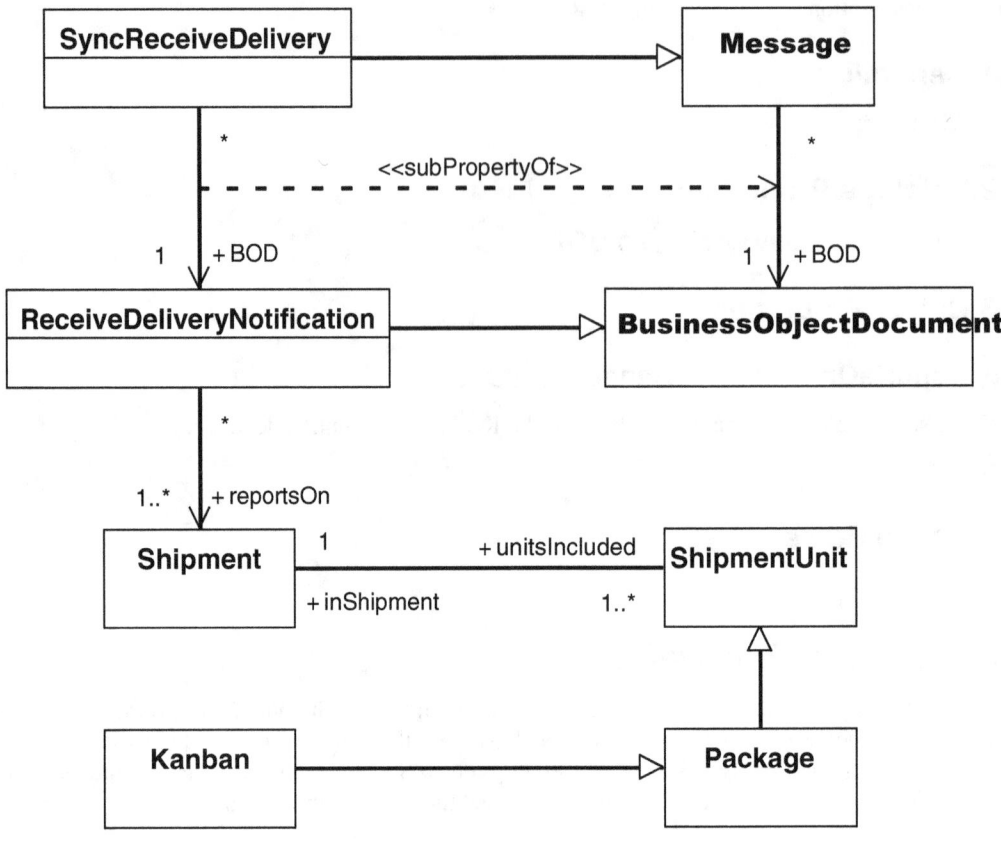

Figure 18   SyncReceiveDelivery Message

12.2.6.1 Generalizations

Message

12.2.6.2 Datatype Properties

none.

12.2.6.3 Object Properties

Property: BOD   **Range:** ReceiveDeliveryNotification

SubpropertyOf: Message.BOD

Definition: The ReceiveDeliveryNotification document that is the payload of the SyncReceiveDelivery message. Specializes Message.BOD.

Cardinality: 1..1

12.2.6.4 Range Roles

none.

12.2.7 Class: ReceiveDeliveryNotification

Definition: A BusinessObjectDocument that shows the state of a Shipment as received at the Customer/ShipTo location. The ReceiveDeliveryNotification includes the following properties of a Shipment as received:

dateReceived, quantityReceived, discrepancies. And it may include additional information on particular ShipmentUnits in the Shipment, such as Kanban number and status.

Note: The presence of a ReceiptDiscrepancy (see 9.2.3) in a ReceiveDeliveryNotification is a signal to the Supplier, or to the supplier's application (System), that an actual Shipment is faulty in some way.

Note: ReceiveDeliveryNotification is depicted in Figure 18.

12.2.7.1 Generalizations

BusinessObjectDocument

12.2.7.2 Datatype Properties

none.

12.2.7.3 Object Properties

Property: reportsOn   **Range: Shipment**

Definition: The Shipment(s) that is the subject of the ReceiveDeliveryNotification document.

Cardinality: 1..*

12.2.7.4 Range Roles

From SyncReceiveDelivery as BOD

12.2.8 Class: SyncShipment

Definition: Also called Advance Shipment Notification. This Message is used by the Supplier to notify the Customer that an authorized Shipment is planned for a specific time period. Depending on the nature of the ShipmentSchedule, the notification may reflect the expected date/time of arrival at the ShipTo facility, the planned or actual date/time of departure from the ShipFrom facility, or the planned or actual window of availability for Customer pickup at the ShipFrom facility.

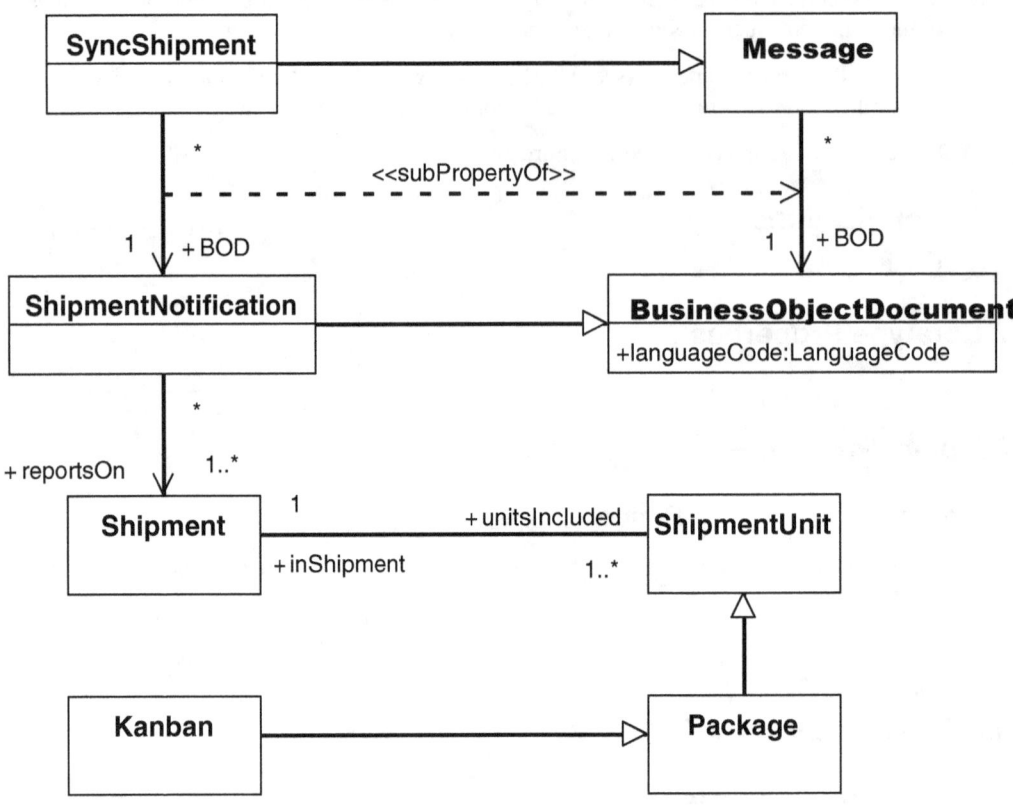

Figure 19 SyncShipment Message

12.2.8.1 Generalizations

Message

12.2.8.2 Datatype Properties

none.

12.2.8.3 Object Properties

Property: BOD   **Range:** ShipmentNotification

SubpropertyOf: Message.BOD

Definition: The (Advance)ShipmentNotification document that is the payload of the SyncShipment message. Specializes Message.BOD.

Cardinality: 1..1

12.2.8.4 Range Roles

none.

12.2.9 Class: ShipmentNotification

Definition: A BusinessObjectDocument that provides advance notice of Shipments to be loaded or en route. It is sent by the Supplier according to the business agreements.

Note: Note: ShipmentNotification is depicted in Figure 19.

12.2.9.1 Generalizations

BusinessObjectDocument

12.2.9.2 Datatype Properties

none.

12.2.9.3 Object Properties

Property: reportsOn **Range: Shipment**

Definition: The Shipment(s) that are the subject of the ShipmentNotification document.

Cardinality: 1..*

12.2.9.4 Range Roles

From SyncShipment as BOD

12.2.10 Class: SyncShipmentSchedule

Definition: A Message sent by the Customer to authorize the shipment of one or more Kanbans to the Customer site, according to an existing KanbanLoop arrangement and ShipmentSchedule.

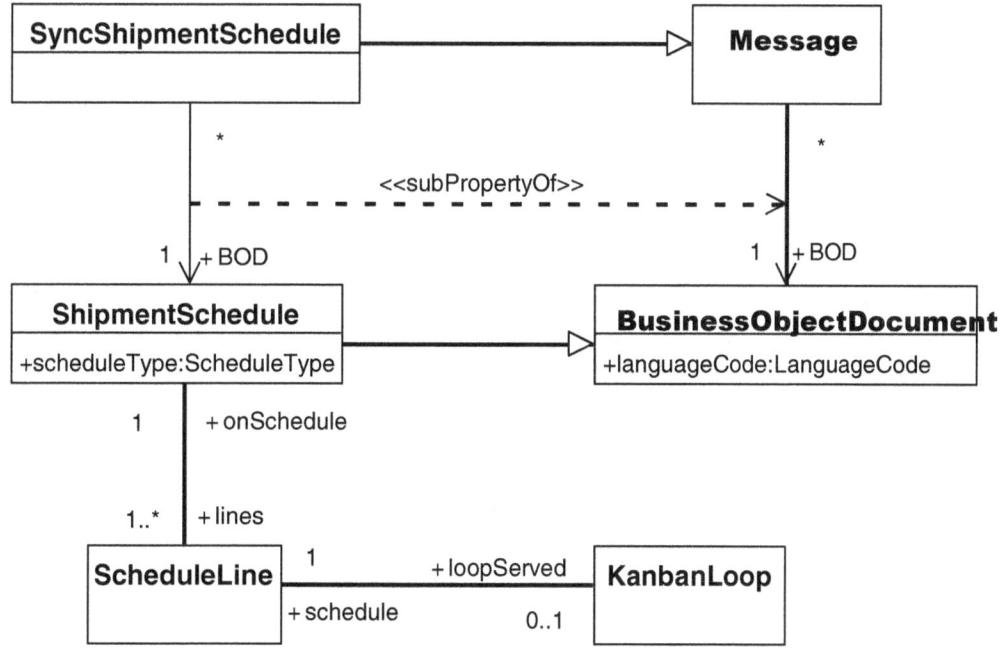

Figure 20   SyncShipmentSchedule Message

12.2.10.1 Generalizations

Message

12.2.10.2 DatatypeProperties

none.

12.2.10.3 ObjectProperties

Property: BOD **Range:** ShipmentSchedule

SubpropertyOf: Message.BOD

Definition: The ShipmentSchedule document that is the payload of the SyncShipmentSchedule message. Specializes Message.BOD.

Cardinality: 1..1

12.2.10.4 RangeRoles

none.

12.3 Instances

none.

References

[1] Automotive Industry Action Group, Publication I-1, *Inventory Visibility and Interoperability – Proof of Concept – Phase I Project Summary*, Southfield, Michigan, March, 2005.

[2] Kupanhy, L., *Classification of JIT techniques and their implications*, in Industrial Engineering, February, 1995 v27 n2 p62-66.

[3] Daconta, M., Obrst, L., and Smith, K., *The Semantic Web: A Guide to the Future of XML, Web Services, and Knowledge Management*, John Wiley & Sons, ISBN 0-471-43257-1, May, 2003.

[4] World Wide Web Consortium (W3C) *OWL Web Ontology Language Reference*, W3C Recommendation 10 February 2004. Available from: http://www.w3.org/TR/owl-ref/ [cited 1 Nov 2006]

[5] Object Management Group (OMG) *Unified Modeling Language (UML)*, version 2.0, July 2005. Available from http://www.omg.org/technology/documents/formal/uml.htm [cited 1 Nov 2006]

[6] United Nations Directories for Electronic Data Interchange for Administration, Commerce and Transport (UN/EDIFACT) Code List 3055 *Responsible Agency Code*, Release D06A, United Nations Economic Commission for Europe, Geneva, 2006.

[7] Hoffman, P., Masinter, L., Zawinski, J., Internet Engineering Task Force (IETF) Request for Comments (RFC) 2368 *The mailto URL scheme*, July 1998. Available from http://www.ietf.org/rfc/rfc2368.txt [cited 1 Nov 2006]

[8] United Nations Directories for Electronic Data Interchange for Administration, Commerce and Transport (UN/EDIFACT) Code List 1001 *Document Name Code*, Release D06A, United Nations Economic Commission for Europe, Geneva, 2006.

[9] Accredited Standards Committee X12, *Electronic Data Interchange* (EDI), version 5030, October 2005.

[10] United Nations Directories for Electronic Data Interchange for Administration, Commerce and Transport (UN/EDIFACT) Code List 3227 *Location function code qualifier*, Release D06A, United Nations Economic Commission for Europe, Geneva, 2006

[11] United Nations Centre for Trade Facilitation and Electronic Business (UN/CEFACT) Recommendation No. 16 *LOCODE - Code for Trade and Transport Locations*, Third Edition, United Nations Economic Commission for Europe, Geneva, 1998.

[12] United Nations Directories for Electronic Data Interchange for Administration, Commerce and Transport (UN/EDIFACT), Release D06A, United Nations Economic Commission for Europe, Geneva, 2006.

[13] ANS MH10.8.2-2002 - *Data Identifier and Application Identifier Standard*, Material Handling Institute of America, Charlotte, North Carolina, 2003.

[14] Automotive Industry Action Group, Publication B-16, *Global Transport Data Identifiers for the Automotive Industry*, Southfield, Michigan, 2004.

[15] ANS MH10.8.3-2002 - *Transfer Data Syntax for High Capacity ADC Media*, Material Handling Institute of America, Charlotte, North Carolina, 2004.

[16] United Nations Directories for Electronic Data Interchange for Administration, Commerce and Transport (UN/EDIFACT) Code List 8053 *Equipment qualifier*, Release D06A, United Nations Economic Commission for Europe, Geneva, 2006.

[17] Open Applications Group, *OAGIS 9.0 - Open Applications Group Integration Specification*, September, 2006. Available from: http://www.openapplications.org/downloads/oagis/loadfrm9.htm [cited 1 Nov 2006].

www.ingramcontent.com/pod-product-compliance
Lightning Source LLC
Chambersburg PA
CBHW081732170526
45167CB00009B/3788